INTERNATIONAL SERIES OF MONOGRAPHS ON PHYSICS

SERIES EDITORS

J. BIRMAN	CITY UNIVERSITY OF NEW YORK
S. F. EDWARDS	UNIVERSITY OF CAMBRIDGE
R. FRIEND	UNIVERSITY OF CAMBRIDGE
M. REES	UNIVERSITY OF CAMBRIDGE
D. SHERRINGTON	UNIVERSITY OF OXFORD
G. VENEZIANO	CERN, GENEVA

INTERNATIONAL SERIES OF MONOGRAPHS ON PHYSICS

160. C. Barrabès, P.A. Hogan: *Advanced general relativity–gravity waves, spinning particles, and black holes*
159. W. Barford: *Electronic and optical properties of conjugated polymers, Second edition*
158. F. Strocchi: *An introduction to non-perturbative foundations of quantum field theory*
157. K.H. Bennemann, J.B. Ketterson: *Novel superfluids, Volume 2*
156. K.H. Bennemann, J.B. Ketterson: *Novel superfluids, Volume 1*
155. C. Kiefer: *Quantum gravity, Third edition*
154. L. Mestel: *Stellar magnetism, Second edition*
153. R. A. Klemm: *Layered superconductors, Volume 1*
152. E.L. Wolf: *Principles of electron tunneling spectroscopy, Second edition*
151. R. Blinc: *Advanced ferroelectricity*
150. L. Berthier, G. Biroli, J.-P. Bouchaud, W. van Saarloos, L. Cipelletti: *Dynamical heterogeneities in glasses, colloids, and granular media*
149. J. Wesson: *Tokamaks, Fourth edition*
148. H. Asada, T. Futamase, P. Hogan: *Equations of motion in general relativity*
147. A. Yaouanc, P. Dalmas de Réotier: *Muon spin rotation, relaxation, and resonance*
146. B. McCoy: *Advanced statistical mechanics*
145. M. Bordag, G.L. Klimchitskaya, U. Mohideen, V.M. Mostepanenko: *Advances in the Casimir effect*
144. T.R. Field: *Electromagnetic scattering from random media*
143. W. Götze: *Complex dynamics of glass-forming liquids-a mode-coupling theory*
142. V.M. Agranovich: *Excitations in organic solids*
141. W.T. Grandy: *Entropy and the time evolution of macroscopic systems*
140. M. Alcubierre: *Introduction to 3+1 numerical relativity*
139. A. L. Ivanov, S. G. Tikhodeev: *Problems of condensed matter physics-quantum coherence phenomena in electron-hole and coupled matter-light systems*
138. I. M. Vardavas, F. W. Taylor: *Radiation and climate*
137. A. F. Borghesani: *Ions and electrons in liquid helium*
135. V. Fortov, I. Iakubov, A. Khrapak: *Physics of strongly coupled plasma*
134. G. Fredrickson: *The equilibrium theory of inhomogeneous polymers*
133. H. Suhl: *Relaxation processes in micromagnetics*
132. J. Terning: *Modern supersymmetry*
131. M. Mariño: *Chern-Simons theory, matrix models, and topological strings*
130. V. Gantmakher: *Electrons and disorder in solids*
129. W. Barford: *Electronic and optical properties of conjugated polymers*
128. R. E. Raab, O. L. de Lange: *Multipole theory in electromagnetism*
127. A. Larkin, A. Varlamov: *Theory of fluctuations in superconductors*
126. P. Goldbart, N. Goldenfeld, D. Sherrington: *Stealing the gold*
125. S. Atzeni, J. Meyer-ter-Vehn: *The physics of inertial fusion*
123. T. Fujimoto: *Plasma spectroscopy*
122. K. Fujikawa, H. Suzuki: *Path integrals and quantum anomalies*
121. T. Giamarchi: *Quantum physics in one dimension*
120. M. Warner, E. Terentjev: *Liquid crystal elastomers*
119. L. Jacak, P. Sitko, K. Wieczorek, A. Wojs: *Quantum Hall systems*
117. G. Volovik: *The Universe in a helium droplet*
116. L. Pitaevskii, S. Stringari: *Bose-Einstein condensation*
115. G. Dissertori, I.G. Knowles, M. Schmelling: *Quantum chromodynamics*
114. B. DeWitt: *The global approach to quantum field theory*
113. J. Zinn-Justin: *Quantum field theory and critical phenomena, Fourth edition*
112. R.M. Mazo: *Brownian motion-fluctuations, dynamics, and applications*
111. H. Nishimori: *Statistical physics of spin glasses and information processing-an introduction*
110. N.B. Kopnin: *Theory of nonequilibrium superconductivity*
109. A. Aharoni: *Introduction to the theory of ferromagnetism, Second edition*
108. R. Dobbs: *Helium three*
107. R. Wigmans: *Calorimetry*
106. J. Kübler: *Theory of itinerant electron magnetism*
105. Y. Kuramoto, Y. Kitaoka: *Dynamics of heavy electrons*
104. D. Bardin, G. Passarino: *The Standard Model in the making*
103. G. C. Branco, L. Lavoura, J.P. Silva: *CP Violation*
102. T. C. Choy: *Effective medium theory*
101. H. Araki *Mathematical theory of quantum fields*
100. L. M. Pismen: *Vortices in nonlinear fields*
99. L. Mestel: *Stellar magnetism*
98. K. H. Bennemann: *Nonlinear optics in metals*
94. S. Chikazumi: *Physics of ferromagnetism*
91. R. A. Berthmann: *Anomalies in quantum field theory*
90. P. K. Gosh: *Ion traps*
87. P. S. Joshi: *Global aspects in gravitation and cosmology*
86. E. R. Pike, S. Sarkar: *The quantum theory of radiation*
83. P. G. de Gennes, J. Prost: *The physics of liquid crystals*
73. M. Doi, S. F. Edwards: *The theory of polymer dynamics*
69. S. Chandrasekhar: *The mathematical theory of black holes*
51. C. Møller: *The theory of relativity*
46. H. E. Stanley: *Introduction to phase transitions and critical phenomena*
32. A. Abragam: *Principles of nuclear magnetism*
27. P. A. M. Dirac: *Principles of quantum mechanics*
23. R. E. Peierls: *Quantum theory of solids*

Advanced General Relativity

*Gravity Waves, Spinning Particles,
and Black Holes*

C. Barrabès
Université de Tours

P. A. Hogan
University College Dublin

OXFORD
UNIVERSITY PRESS

Great Clarendon Street, Oxford, OX2 6DP,
United Kingdom

Oxford University Press is a department of the University of Oxford.
It furthers the University's objective of excellence in research, scholarship,
and education by publishing worldwide. Oxford is a registered trade mark of
Oxford University Press in the UK and in certain other countries

© Claude Barrabès, Peter A. Hogan 2013

The moral rights of the authors have been asserted

First Edition published in 2013

Impression: 1

All rights reserved. No part of this publication may be reproduced, stored in
a retrieval system, or transmitted, in any form or by any means, without the
prior permission in writing of Oxford University Press, or as expressly permitted
by law, by licence or under terms agreed with the appropriate reprographics
rights organization. Enquiries concerning reproduction outside the scope of the
above should be sent to the Rights Department, Oxford University Press, at the
address above

You must not circulate this work in any other form
and you must impose this same condition on any acquirer

Published in the United States of America by Oxford University Press
198 Madison Avenue, New York, NY 10016, United States of America

British Library Cataloguing in Publication Data

Data available

Library of Congress Control Number: 2013938191

ISBN 978-0-19-968069-6

Printed and bound by
CPI Group (UK) Ltd, Croydon, CR0 4YY

Links to third party websites are provided by Oxford in good faith and
for information only. Oxford disclaims any responsibility for the materials
contained in any third party website referenced in this work.

Preface

This book is aimed at students making the transition from a final year undergraduate course on general relativity, based on one of the many introductory texts, to a specialized subfield of general relativity covered by an existing research monograph. We present a variety of topics under the general headings of gravitational waves *in vacuo* and in a cosmological setting, equations of motion, and black holes, all having a clear physical relevance and a strong emphasis on space–time geometry. The topic of black hole physics is particularly extensive and so a selection of classical and quantum aspects, with suitable introduction, has been made with a view to complementing the recent text, *Introduction to Black Hole Physics*, by Frolov and Zelnikov (2011). Each chapter in this book could be used as a basis for an advanced undergraduate or early postgraduate project since our intention is to whet the appetite of readers who are exploring avenues into research in general relativity and who have already accumulated the required technical knowledge. To be more specific we expect the reader to have completed an introductory course on general relativity at the level of *Introducing Einstein's Relativity* by Ray d'Inverno (1992) and then to have supplemented this material with additional techniques by individual study or in a taught MSc programme. The additional technical knowledge required involves the Cartan calculus, the tetrad formalism including aspects of the Newman–Penrose formalism, the Ehlers–Sachs theory of null geodesic congruences, and the Petrov classification of gravitational fields, all of which are treated clearly and economically by S. Chandrasekhar (1983) in the mathematical preliminaries for *The Mathematical Theory of Black Holes*. For the geometry underpinning the cosmology treated here it would be hard to surpass the lecture notes of G. F. R. Ellis (1971). The topics covered in this book are not individual applications of any one of these techniques. Our attitude to the techniques is to regard them as available to us in whole or in part (mainly in part) as each situation demands. The presentation of each chapter is research monograph style rather than textbook style in order to impress on interested students the need to present their research in a clear and concise format. Our hope is that students with our taste in general relativity will find a treasure trove here.

Acknowledgements

We thank the Université de Tours, the Centre National de la Recherche Scientifique (CNRS), and the Ministère des Affaires Étrangères for supporting our collaboration over many years. The work described in Chapter 3 on spinning particles was carried out in collaboration with Mr. Shinpei Ogawa and we are grateful to him for his decisive contribution. We thank Dr. Loic Villain for his expert help in the production of Chapter 5.

Contents

1	**Minkowskian space–time**	1
	1.1 Lorentz tansformations	1
	1.2 Non-singular and singular Lorentz transformations	3
	1.3 Infinitesimal Lorentz transformations	6
	1.4 Geometrical construction of a gravitational wave	8
2	**Plane gravitational waves**	10
	2.1 From linear approximation to colliding waves	10
	2.2 Electromagnetic shock waves	17
	2.3 Gravitational shock waves	21
	2.4 High-frequency gravity waves	24
3	**Equations of motion**	28
	3.1 Motivation	28
	3.2 Example of a background space–time	33
	3.3 Equations of motion of a Reissner–Nordström particle in first approximation	37
	3.4 Background space–time for a Kerr particle	39
	3.5 Equations of motion of a Kerr particle in first approximation	45
	3.6 Spinning test particles	53
4	**Inhomogeneous aspects of cosmology**	56
	4.1 Plane-fronted gravitational waves with a cosmological constant	56
	4.2 Perturbations of isotropic cosmologies	60
	4.3 Gravitational waves	64
	4.4 Cosmic background radiation	73
5	**Black holes**	79
	5.1 Introduction: Basic properties of black holes	79
	5.2 Collapsing null shells and trapped surface formation	86
	5.3 Scattering properties of high-speed Kerr black holes	91
	5.4 Inside the black hole	103
	5.5 Metric fluctuations and Hawking radiation	110
6	**Higher dimensional black holes**	118
	6.1 Brief outline of D-dimensional black holes	119
	6.2 Gibbons–Penrose isoperimetric inequality and the hoop conjecture in D dimensions	122
	6.3 Light-like boost of higher dimensional black holes	124

Appendix A Notation	131
Appendix B Transport law for k along $r=0$	133
Appendix C Some useful scalar products	135
References	137
Index	142

1
Minkowskian space–time

Space–time geometry was introduced into relativity theory in the classical paper by Minkowski (1909) [translated by Lorentz (1923)] and so we begin our presentation from the space–time viewpoint on general relativity by considering some aspects of Minkowskian space–time. We discuss non-singular and singular Lorentz transformations, infinitesimal Lorentz transformations, and exploit the similarities of the latter to electromagnetic fields. An elegant presentation, which has influenced us, is that of Trautman *et al.* (1965). A natural generalization of the Lorentz transformations leads to the geometrical construction of a gravitational wave.

1.1 Lorentz tansformations

The line-element of Minkowskian space–time, in rectangular Cartesian coordinates and time, is given by

$$ds^2 = dx^2 + dy^2 + dz^2 - dt^2 = \eta_{ij}\, dx^i\, dx^j. \tag{1.1}$$

Here $x^i = (x, y, z, t)$ for $i = 1, 2, 3, 4$, respectively, and we use units in which the speed of light in vacuum is $c = 1$. In general Latin indices will take values 1, 2, 3, 4, the Einstein summation convention will apply, and $(\eta_{ij}) = \mathrm{diag}(1, 1, 1, -1)$ are the components of the Minkowskian metric tensor in coordinates x^i with $(\eta^{ij}) = (\eta_{ij})^{-1} = \mathrm{diag}(1, 1, 1, -1)$ and Latin indices are raised and lowered as usual with η^{ij} and η_{ij}, respectively.

There is a one-to-one correspondence between points of Minkowskian space–time with position vectors $\mathbf{x} = (x, y, z, t)$ and 2×2 Hermitean matrices. Given a point $\mathbf{x}$ we write the corresponding 2×2 Hermitean matrix in the form

$$A(\mathbf{x}) = \begin{pmatrix} -z + t & x + iy \\ x - iy & z + t \end{pmatrix}. \tag{1.2}$$

The standard way of demonstrating the action of the proper, orthochronous Lorentz transformations on Minkowskian space–time is to recognize that if U is a 2×2 matrix with complex entries and unit determinant (unimodular), and thus involving six real parameters, then the 2×2 matrix $UA(\mathbf{x})U^\dagger$, where $U^\dagger$ denotes the Hermitean conjugate of U, is itself a 2×2 Hermitean matrix and hence there exists a point $\mathbf{x}' = (x', y', z', t')$ of Minkowskian space–time for which

$$A(\mathbf{x}') = UA(\mathbf{x})U^\dagger. \tag{1.3}$$

2 Minkowskian space–time

Calculating the determinants of both sides of this matrix equation immediately results in
$$-x'^2 - y'^2 - z'^2 + t'^2 = -x^2 - y^2 - z^2 + t^2. \tag{1.4}$$

Hence the transformation from **x** to **x'**, and vice versa, implicit in (1.3) is a Lorentz transformation which on more detailed examination is revealed to be proper (orientation preserving) and orthochronous (preserving the time direction). We will henceforth refer to such transformations simply as Lorentz transformations. On account of the quadratic dependence on U in (1.3) it is clear that there are *two* unimodular matrices $\pm U$ corresponding to each Lorentz transformation. The matrices U are elements of the group $SL(2,\mathbf{C})$ while the transformations from **x** to **x'**, and vice versa, constitute the six-real-parameter proper, orthochronous Lorentz group.

Given a unimodular matrix U it is straightforward to calculate explicitly the corresponding Lorentz transformation from (1.3). It is not quite so straightforward to calculate the two unimodular matrices corresponding to a given Lorentz transformation and so a little practice is useful. The reader can check that the diagonal matrices

$$U = \pm \begin{pmatrix} e^{-i\phi/2} & 0 \\ 0 & e^{i\phi/2} \end{pmatrix}, \tag{1.5}$$

correspond to the spatial rotation

$$x' = x \cos\phi + y \sin\phi, \tag{1.6}$$
$$y' = -x \sin\phi + y \cos\phi, \tag{1.7}$$
$$z' = z, \tag{1.8}$$
$$t' = t, \tag{1.9}$$

whereas the matrices

$$U = \pm \begin{pmatrix} \cos\frac{\theta}{2} & \sin\frac{\theta}{2} \\ -\sin\frac{\theta}{2} & \cos\frac{\theta}{2} \end{pmatrix}, \tag{1.10}$$

correspond to the spatial rotation

$$x' = x \cos\theta + z \sin\theta, \tag{1.11}$$
$$y' = y, \tag{1.12}$$
$$z' = -x \sin\theta + z \cos\theta, \tag{1.13}$$
$$t' = t. \tag{1.14}$$

Also the diagonal matrices

$$U = \pm \begin{pmatrix} \left(\frac{1-v}{1+v}\right)^{-1/4} & 0 \\ 0 & \left(\frac{1-v}{1+v}\right)^{1/4} \end{pmatrix}, \tag{1.15}$$

correspond to the Lorentz boost

$$x' = x, \tag{1.16}$$
$$y' = y, \tag{1.17}$$

$$z' = \gamma(v)(z - vt), \tag{1.18}$$
$$t' = \gamma(v)(t - vz), \tag{1.19}$$

where $\gamma(v) = (1-v^2)^{-1/2}$ is the Lorentz factor, and the matrices

$$U = \pm \frac{1}{\sqrt{2}} \begin{pmatrix} \sqrt{\gamma(v)+1} & -\sqrt{\gamma(v)-1} \\ -\sqrt{\gamma(v)-1} & \sqrt{\gamma(v)+1} \end{pmatrix}, \tag{1.20}$$

correspond to the Lorentz boost

$$x' = \gamma(v)(x - vt), \tag{1.21}$$
$$y' = y, \tag{1.22}$$
$$z' = z, \tag{1.23}$$
$$t' = \gamma(v)(t - vx). \tag{1.24}$$

These one-parameter subgroups of the proper, orthochronous Lorentz group are useful for illustrative purposes below. In the latter two examples the primed frame of reference is moving with constant 3-velocity $v < 1$ in the z-direction relative to the unprimed frame in the case of (1.16)–(1.19) and in the x-direction relative to the unprimed frame in the case of (1.21)–(1.24).

1.2 Non-singular and singular Lorentz transformations

The position vector of a point on the future null-cone with vertex at the origin $(0,0,0,0)$ of the coordinates x^i in Minkowskian space–time is given by

$$\mathbf{x} = (x, y, z, t) \text{ with } t > 0 \text{ and } x^2 + y^2 + z^2 = t^2. \tag{1.25}$$

This vector can be written in parametrized form as

$$\mathbf{x} = t(\sin\theta\cos\phi, \sin\theta\sin\phi, \cos\theta, 1), \tag{1.26}$$

with the polar angles θ, ϕ having the usual ranges $0 \leq \theta \leq \pi$ and $0 \leq \phi < 2\pi$. Thus a null *direction* on the future null-cone (1.25) is specified by the angles θ, ϕ. It is convenient to use the complex number

$$\zeta = e^{i\phi}\cotan\frac{\theta}{2}, \tag{1.27}$$

in place of the polar angles. The real and imaginary parts of this complex number specify the image in the equatorial plane of the stereographic projection from the north pole (corresponding to $\theta = 0$) of a point on the unit 2-sphere. Thus (1.26) takes the form

$$\mathbf{x} = t\left(\frac{\bar\zeta + \zeta}{1 + \zeta\bar\zeta}, \frac{i(\bar\zeta - \zeta)}{1 + \zeta\bar\zeta}, \frac{\zeta\bar\zeta - 1}{\zeta\bar\zeta + 1}, 1\right), \tag{1.28}$$

with the bar denoting complex conjugation. Hence the points of the extended complex plane (the complex plane including the point at infinity $\zeta = \infty$, with the latter corresponding to $\theta = 0$) are required to specify all of the null directions on the future null-cone (1.25). Clearly the point at infinity of the extended complex plane specifies

4 Minkowskian space–time

the generator $t = z$ of the null-cone (1.25). Under a proper, orthochronous Lorentz transformation, (1.28) is transformed to another null position vector

$$\mathbf{x}' = t' \left(\frac{\bar{\zeta}' + \zeta'}{1 + \zeta'\bar{\zeta}'}, \frac{i(\bar{\zeta}' - \zeta')}{1 + \zeta'\bar{\zeta}'}, \frac{\zeta'\bar{\zeta}' - 1}{\zeta'\bar{\zeta}' + 1}, 1 \right), \tag{1.29}$$

whose *direction* is specified by the complex number ζ'. Now

$$A(\mathbf{x}) = \frac{2t}{1 + \zeta\bar{\zeta}} \begin{pmatrix} 1 & \zeta \\ \bar{\zeta} & \zeta\bar{\zeta} \end{pmatrix}, \tag{1.30}$$

with a similar equation for $A(\mathbf{x}')$. The relationship between ζ' and ζ will tell us how null directions on the null-cone are transformed under the Lorentz transformations under consideration. This information is readily obtained by substituting (1.30) and the corresponding expression for $A(\mathbf{x}')$ into (1.3) for any

$$U = \begin{pmatrix} \alpha_0 & \beta_0 \\ \gamma_0 & \delta_0 \end{pmatrix}, \tag{1.31}$$

where $\alpha_0, \beta_0, \gamma_0, \delta_0$ are complex numbers satisfying $\alpha_0\delta_0 - \beta_0\gamma_0 = 1$. Straightforward algebra reveals that

$$\zeta' = \frac{\bar{\gamma}_0 + \bar{\delta}_0 \zeta}{\bar{\alpha}_0 + \bar{\beta}_0 \zeta}. \tag{1.32}$$

Such transformations constitute the fractional linear group of transformations of the extended complex plane. We have noted that there are two matrices U (differing only in sign) corresponding to each Lorentz transformation whereas it is clear from (1.32) that there is a one-to-one correspondence between proper, orthochronous Lorentz transformations and fractional linear transformations. In fact these latter two groups of transformations are isomorphic.

Fixed points of the transformation (1.32) correspond to null directions left invariant by Lorentz transformations. For a given Lorentz transformation the fixed points ζ are given by the roots of the quadratic equation

$$\bar{\beta}_0 \zeta^2 + (\bar{\delta}_0 - \bar{\alpha}_0)\zeta - \bar{\gamma}_0 = 0, \tag{1.33}$$

over the field of complex numbers. Hence in general a proper, orthochronous Lorentz transformation leaves *two* null directions on the null-cone invariant. Such transformations are called *non-singular*. However it is obviously possible to have Lorentz transformations for which the roots of the quadratic equation (1.33) are equal. In these cases only *one* null direction on the null-cone is left invariant. Such transformations are called *singular* Lorentz transformations or *null rotations*.

The fixed points of the transformation (1.32) corresponding to the non-diagonal case (1.10) above are $\zeta = \pm i$ and the corresponding invariant null directions, obtained from (1.28), are tangent to the null geodesic generators $y = \pm t$ of the null-cone with vertex $(0, 0, 0, 0)$. The fixed points corresponding to (1.20) are $\zeta = \pm 1$ and the corresponding invariant null directions are tangent to the lines $x = \pm t$. The diagonal cases (1.5) and (1.15) both have $\beta_0 = 0 = \gamma_0$ and $\alpha_0 \neq \delta_0$. In this case we see that (1.33) has the solution $\zeta = 0$. Also rewriting (1.33) in the form

$$\bar{\beta}_0 + (\bar{\delta}_0 - \bar{\alpha}_0)\zeta^{-1} - \bar{\gamma}_0 \zeta^{-2} = 0, \tag{1.34}$$

it follows that $\zeta = \infty$ when $\beta_0 = 0 = \gamma_0$ and $\alpha_0 \neq \delta_0$. Hence the invariant null directions in the cases (1.5) and (1.15) are tangent to the lines $z = \pm t$. Consequently all four examples in the previous section are non-singular Lorentz transformations. An example of a singular Lorentz transformation is

$$x' + iy' = x + iy + w(t+z), \tag{1.35}$$

$$t' - z' = t - z + w\bar{w}(t+z) + w(x - iy) + \bar{w}(x + iy), \tag{1.36}$$

$$t' + z' = t + z, \tag{1.37}$$

where $w \neq 0$ ($w = 0$ for the identity transformation) is a complex number with complex conjugate $\bar{w}$. In this form it is easy to check that (1.4) is satisfied. The corresponding matrices U are found to be

$$U = \pm \begin{pmatrix} 1 & w \\ 0 & 1 \end{pmatrix}. \tag{1.38}$$

The unique fixed point of (1.32) is thus $\zeta = 0$ and the corresponding invariant null direction is tangent to the line $z = -t$. The transformations (1.35)–(1.37) constitute an Abelian, two-(real)-parameter subgroup of the proper, orthochronous Lorentz group. It is interesting to transform (1.35)–(1.37) from the coordinates (x, y, z, t) to the coordinates (ξ, η, r, u) via

$$\xi = \frac{x}{z+t}, \tag{1.39}$$

$$\eta = \frac{y}{z+t}, \tag{1.40}$$

$$r = z + t, \tag{1.41}$$

$$u = -\frac{1}{2}(z - t) - \frac{1}{2}\frac{(x^2 + y^2)}{z+t}. \tag{1.42}$$

Now (1.35)–(1.37) takes the simpler form

$$\xi' + i\eta' = \xi + i\eta + w \, , \, r' = r \, , \, u' = u. \tag{1.43}$$

Even more revealing is to write the Minkowskian line-element (1.1) in the coordinates (ξ, η, r, u). This results in

$$ds^2 = r^2(d\xi^2 + d\eta^2) - 2\, du\, dr, \tag{1.44}$$

which was first given by Ivor Robinson (see Rindler and Trautman (1987)) as a byproduct of his study of the Schwarzschild line-element in the limit of the mass $m \to +\infty$. It is immediate from (1.44) that $r = 0$ is a null geodesic with u an affine parameter along it and that (1.43) is a Lorentz transformation which leaves this null geodesic invariant. This observation by Robinson is of great significance in the history of singular Lorentz transformations [see Synge (1965), p.viii].

Robinson's interesting limit of the Schwarzschild solution referred to above is obtained by first writing the Schwarzschild line-element in the form (a more common form of the Schwarzschild line-element can be found in (3.1) below)

$$ds^2 = \frac{r^2(d\xi^2 + d\eta^2)}{\cosh^2 \lambda \xi} - 2\, du\, dr - \left(\lambda^2 - \frac{2}{r}\right) du^2, \tag{1.45}$$

with

$$\lambda = m^{-1/3}, \tag{1.46}$$

and then taking the limit $\lambda \to 0$ (equivalently $m \to +\infty$) to arrive at the line-element

$$ds^2 = r^2(d\xi^2 + d\eta^2) - 2\,du\,dr + \frac{2}{r}\,du^2. \tag{1.47}$$

This is the flat space–time line-element (1.44) with the addition of a term (the final term) which is singular at $r = 0$. The reader may like to show that, with a suitable coordinate transformation, it can be put in the vacuum Kasner (1925) form:

$$ds^2 = T^{4/3}(dX^2 + dY^2) + T^{-2/3}dZ^2 - dT^2. \tag{1.48}$$

1.3 Infinitesimal Lorentz transformations

A convenient way to introduce infinitesimal Lorentz transformations in the formalism above is via the approximately unimodular matrices

$$U = \pm \begin{pmatrix} 1 + \epsilon(a_1 + ia_2) + O(\epsilon^2) & \epsilon(b_1 + ib_2) + O(\epsilon^2) \\ \epsilon(c_1 + ic_2) + O(\epsilon^2) & 1 - \epsilon(a_1 + ia_2) + O(\epsilon^2) \end{pmatrix}. \tag{1.49}$$

Here ϵ is a small real parameter controlling the approximation where we neglect $O(\epsilon^2)$ terms but keep a note of their presence. Also $a_1, a_2, b_1, b_2, c_1, c_2$ are six real numbers. Now using (1.3) the corresponding infinitesimal Lorentz transformation can be written in the form (for $i = 1, 2, 3, 4$)

$$x'^i = x^i + \epsilon L^i_j x^j + O(\epsilon^2), \tag{1.50}$$

where $x'^i = (x', y', z', t')$ and

$$(L^i_j) = \begin{pmatrix} 0 & -2a_2 & b_1 - c_1 & b_1 + c_1 \\ 2a_2 & 0 & b_2 + c_2 & b_2 - c_2 \\ -(b_1 - c_1) & -(b_2 + c_2) & 0 & -2a_1 \\ b_1 + c_1 & b_2 - c_2 & -2a_1 & 0 \end{pmatrix}, \tag{1.51}$$

with the upper index on L indicating the rows and the lower index indicating the columns in this matrix. Let $x^i = x^i(s)$ be a time-like world line in Minkowskian space–time, with s the arc length or proper-time along it. Let $u^i = dx^i/ds$ and then $\eta_{ij} u^i u^j = u_i u^i = -1$ for all s. Thus u^i is the 4-velocity of the particle with world line $x^i = x^i(s)$. Since $u_i u^i$ is conserved along this world line it follows that $u^i(s_1)$ and $u^i(s_2)$, for $s_1 \neq s_2$, must be related by a Lorentz transformation. In particular $u^i(s + \epsilon)$ and $u^i(s)$, for small ϵ, must be related by an infinitesimal Lorentz transformation for which the entries in the matrix (1.51) depend upon s. Thus

$$u^i(s + \epsilon) = u^i(s) + \epsilon L^i_j(s) u^j(s) + O(\epsilon^2). \tag{1.52}$$

Dividing by ϵ and taking the limit $\epsilon \to 0$ yields the propagation law for $u^i(s)$ along the world line:
$$\frac{du^i}{ds} = L^i_j u^j. \tag{1.53}$$

As usual we can write the components of the 4-velocity u^i in terms of the 3-velocity $\vec{u}$ of the particle in the form
$$u^i = \gamma(u)\,(\vec{u},1) \text{ where } \vec{u} = \left(\frac{dx}{dt}, \frac{dy}{dt}, \frac{dz}{dt}\right) \text{ and } \gamma(u) = (1-u^2)^{-1/2}, \tag{1.54}$$

with $u^2 = \vec{u}\cdot\vec{u}$. Hence (1.53) can be written more explicitly as
$$\frac{d}{dt}\left(\gamma(u)\frac{dx^\alpha}{dt}\right) = L^\alpha_\beta \frac{dx^\beta}{dt} + L^\alpha_4, \tag{1.55}$$

$$\frac{d}{dt}\gamma(u) = L^4_\beta \frac{dx^\beta}{dt}, \tag{1.56}$$

where Greek indices take values 1, 2, 3 with the Einstein summation convention continuing to apply and where we have used the fact that $L^4_4 = 0$ from (1.51). Defining the 3-vectors
$$\vec{U} = (b_1 + c_1, b_2 - c_2, -2a_1), \tag{1.57}$$
$$\vec{W} = (b_2 + c_2, -b_1 + c_1, -2a_2), \tag{1.58}$$

we can rewrite (1.55) and (1.56) in 3-vector notation as
$$\frac{d}{dt}(\gamma(u)\vec{u}) = \vec{u} \times \vec{W} + \vec{U}, \tag{1.59}$$

$$\frac{d\gamma(u)}{dt} = \vec{u}\cdot\vec{U}. \tag{1.60}$$

It is easy to see, by taking the scalar product of (1.59) with $\vec{u}$, that (1.60) is a consequence of (1.59). The 3-velocity dependence of the 3-force on the right-hand side of (1.59) is identical to the 3-velocity dependence of the classical Lorentz 3-force acting on a charged particle moving in an electromagnetic field. We can make this connection more explicit by writing
$$\vec{U} = \frac{q}{m}\vec{E} = \frac{q}{m}\left(E^1, E^2, E^3\right), \tag{1.61}$$
$$\vec{W} = \frac{q}{m}\vec{B} = \frac{q}{m}\left(B^1, B^2, B^3\right). \tag{1.62}$$

Now (1.59) becomes
$$m\frac{d}{dt}(\gamma(u)\vec{u}) = q\left(\vec{u}\times\vec{B} + \vec{E}\right), \tag{1.63}$$

which is identical to the equations of motion of a charge q of mass m moving in an electric field $\vec{E}$ and a magnetic field $\vec{B}$. The matrix (1.51) can be written
$$(L^i_j) = \frac{q}{m}\begin{pmatrix} 0 & B^3 & -B^2 & E^1 \\ -B^3 & 0 & B^1 & E^2 \\ B^2 & -B^1 & 0 & E^3 \\ E^1 & E^2 & E^3 & 0 \end{pmatrix}. \tag{1.64}$$

We note that if we neglect $O(\epsilon)$ terms then with U given by (1.49) the fixed null directions of the infinitesimal Lorentz transformations generating the transport of u^i along the time-like world line are given by $\zeta(s)$ such that

$$(b_1 - ib_2)\zeta^2 - 2(a_1 - ia_2)\zeta - (c_1 - ic_2) = 0. \tag{1.65}$$

In terms of the vectors $\vec{E}$ and $\vec{B}$ we have (ignoring a factor of q/m which is unnecessary here)

$$a_1 + ia_2 = -\frac{1}{2}\left(E^3 + iB^3\right), \tag{1.66}$$

$$b_1 + ib_2 = \frac{1}{2}\left(E^1 + iE^2\right) + \frac{i}{2}\left(B^1 + iB^2\right), \tag{1.67}$$

$$c_1 + ic_2 = \frac{1}{2}\left(E^1 - iE^2\right) + \frac{i}{2}\left(B^1 - iB^2\right). \tag{1.68}$$

Hence we see that the roots of (1.65) are equal if and only if

$$|\vec{E}|^2 - |\vec{B}|^2 = 0 \quad \text{and} \quad \vec{E} \cdot \vec{B} = 0, \tag{1.69}$$

indicating that the time-like world line of the charged mass is generated by a succession of infinitesimal *singular* Lorentz transformations provided the charge is moving in a purely radiative electromagnetic field. Further information, such as the transformation laws for $\vec{E}$ and $\vec{B}$ under the Lorentz boost (1.21)–(1.24), can be deduced from the construction given here.

1.4 Geometrical construction of a gravitational wave

With $\xi + i\eta = \sqrt{2}\,Z$ we can rewrite (1.39)–(1.42) equivalently as

$$x + iy = r\sqrt{2}\,Z,\quad x - iy = r\sqrt{2}\,\bar{Z},\quad z + t = r \quad \text{and} \quad -z + t = 2u + 2r\,Z\bar{Z}, \tag{1.70}$$

with the bar as usual denoting complex conjugation. Thus the line-element (1.44) takes the form

$$ds^2 = 2\,r^2\,dZ\,d\bar{Z} - 2\,du\,dr. \tag{1.71}$$

It is easy to check now that $u = $ constant are null cones with vertices on the null geodesic $r = 0$. When the proper, orthochronous Lorentz transformation (1.3), with U given by (1.31), is written in terms of the coordinates $Z, \bar{Z}, r, u$ it can be simplified to read

$$r' = 2\gamma_0\bar{\gamma}_0 u + r\,|\sqrt{2}\,\gamma_0\,Z + \delta_0|^2, \tag{1.72}$$

$$r'\,Z' = \sqrt{2}\,\alpha_0\,\bar{\gamma}_0\,u + r\left(\sqrt{2}\,\bar{\gamma}_0\,\bar{Z} + \bar{\delta}_0\right)\left(\frac{1}{\sqrt{2}}\beta_0 + \alpha_0\,Z\right), \tag{1.73}$$

$$u' = \frac{u\,r}{2\gamma_0\,\bar{\gamma}_0\,u + r\,|\sqrt{2}\,\gamma_0\,Z + \delta_0|^2}. \tag{1.74}$$

As a first step in generalizing this transformation let

$$f(Z) = \frac{\frac{1}{\sqrt{2}}\beta_0 + \alpha_0\,Z}{\delta_0 + \sqrt{2}\,\gamma_0\,Z}, \tag{1.75}$$

Geometrical construction of a gravitational wave

and, denoting the derivative of this with a prime, we can write (1.72)–(1.74) in the form (Hogan, 1994)

$$r' = \frac{r}{|f'|}\left(1 + \frac{u}{4r}\left|\frac{f''}{f'}\right|^2\right), \tag{1.76}$$

$$u' = u|f'|\left(1 + \frac{u}{4r}\left|\frac{f''}{f'}\right|^2\right)^{-1}, \tag{1.77}$$

$$Z' = f(Z) - \frac{u}{2r}\frac{f'\bar{f}''}{\bar{f}'}\left(1 + \frac{u}{4r}\left|\frac{f''}{f'}\right|^2\right)^{-1}. \tag{1.78}$$

Now assume that $f(Z)$ is an *arbitrary* analytic function and calculate the line-element from

$$ds^2 = 2\,r'^2\,dZ'\,d\bar{Z}' - 2\,du'\,dr',$$
$$= 2\,r^2\left|dZ - \frac{u}{2r}\bar{H}(\bar{Z})\,d\bar{Z}\right|^2 - 2\,du\,dr, \tag{1.79}$$

where

$$H(Z) = \frac{f'''}{f'} - \frac{3}{2}\left(\frac{f''}{f'}\right)^2. \tag{1.80}$$

The analytic function (1.80) vanishes if and only if the function $f(Z)$ is fractional linear as in (1.75). Penrose's (1972) 'spherical' impulsive gravitational wave having as history in space–time the future null-cone $u = 0$ is obtained from this simply by replacing the coefficient u of $\bar{H}$ in (1.79) by $u\,\vartheta(u)$ where $\vartheta(u)$ is the Heaviside step function which is unity for $u > 0$ and vanishes for $u < 0$. The Ricci tensor, calculated with the metric tensor given via the line-element (1.79) with u replaced by $u\,\vartheta(u)$, vanishes, indicating that we have a vacuum space–time in particular *on* $u = 0$ (the space–time is of course Minkowskian for $u > 0$ and for $u < 0$). There is only one Newman–Penrose component of the Riemann curvature tensor and it is

$$\Psi_4 = \frac{1}{2r}H(Z)\,\delta(u). \tag{1.81}$$

This fact indicates that the Riemann tensor is type N (radiative) in the Petrov classification with $\partial/\partial r$ as degenerate principal null direction. The profile is a Dirac delta function which is singular on the history of the wave $u = 0$ and the field is also singular on the null geodesic $r = 0$ which is a generator of the null-cone $u = 0$. Thus the wavefront has a singular point on it and so the wave is not strictly spherical. Further details including the construction of this wave using a 'cut and paste' approach can be found in Penrose (1972) and Barrabès and Hogan (2003b) and references therein.

2
Plane gravitational waves

Most introductory texts on general relativity present plane gravitational waves of arbitrary profile as solutions of Einstein's vacuum field equations in the linear approximation. Starting in this way we lead the reader through a sequence of gauge transformations to a metric tensor which is an exact solution of the vacuum field equations. We specialize to impulsive plane gravitational waves having a Dirac delta function profile and then give a simple derivation of the Khan–Penrose solution of Einstein's vacuum field equations describing the gravitational field following the head-on collision of such plane waves. By comparison the head-on collision of *electromagnetic* shock waves, having a Heaviside step function profile, leading to the Bell–Szekeres solution of the vacuum Einstein–Maxwell field equations, is described and the corresponding solution of Einstein's field equations for colliding *gravitational* shock waves is also given. Finally we consider high-frequency gravitational waves propagating in a vacuum and examine the similarities between plane and approximately spherical fronted waves.

2.1 From linear approximation to colliding waves

In order to establish notation for this very familiar topic we begin by writing the metric tensor components as small perturbations of the Minkowskian metric tensor thus:

$$g_{ij} = \eta_{ij} + \gamma_{ij}, \qquad (2.1)$$

where η_{ij} is given via (1.1) and $\gamma_{ij} = \gamma_{ji}$ with $\gamma_{ij} = \gamma_{ij}(x, y, z, t)$. We can consider γ_{ij} as the components of a tensor field on Minkowskian space–time with metric tensor having components η_{ij}. Now define the star conjugate of γ_{ij} by

$$\gamma^*_{ij} = \gamma_{ij} - \frac{1}{2}\eta_{ij}\gamma, \qquad (2.2)$$

with $\gamma = \eta^{ij}\gamma_{ij}$. If γ^*_{ij} satisfies the *coordinate conditions*

$$\eta^{jk}\gamma^*_{ij,k} = 0, \qquad (2.3)$$

with the comma denoting partial differentiation with respect to x^k, then Einstein's vacuum field equations in the linear approximation reduce to the 4-dimensional wave equation

$$\left(\frac{\partial^2}{\partial x^2} + \frac{\partial^2}{\partial y^2} + \frac{\partial^2}{\partial z^2} - \frac{\partial^2}{\partial t^2}\right)\gamma_{ij}^* = 0. \tag{2.4}$$

The corresponding gravitational field is described by the linearized Riemann tensor with components

$$R_{ijkm}(\gamma) = \frac{1}{2}\left(\gamma_{im,jk} + \gamma_{jk,im} - \gamma_{ik,jm} - \gamma_{jm,ik}\right). \tag{2.5}$$

If $\xi^i(x, y, z, t)$ are the components of a vector field on Minkowskian space–time, with $\xi_i = \eta_{ij}\xi^j$, and if ξ^i satisfies the wave equation

$$\left(\frac{\partial^2}{\partial x^2} + \frac{\partial^2}{\partial y^2} + \frac{\partial^2}{\partial z^2} - \frac{\partial^2}{\partial t^2}\right)\xi^i = 0, \tag{2.6}$$

then the equations (2.3), (2.4), and (2.5) are invariant under the transformation

$$\gamma_{ij} \to \bar{\gamma}_{ij} = \gamma_{ij} - \xi_{i,j} - \xi_{j,i}. \tag{2.7}$$

Since in particular this means that $R_{ijkm}(\gamma) = R_{ijkm}(\bar{\gamma})$ we shall refer to (2.7) as a *gauge transformation*.

To obtain plane wave solutions with arbitrary profile travelling in the positive z-direction (say) let $u = z - t = k_i x^i$ and take

$$\gamma_{ij}^* = \gamma_{ij}^*(u) , \quad \gamma_{ij}^* k^j = 0. \tag{2.8}$$

Now since $\gamma_{ij,k}^* = \dot{\gamma}_{ij}^* k_k$, with the dot denoting differentiation with respect to u, and $\eta_{ij}k^i k^j = k_i k^i = 0$ it is straightforward to see that (2.3) and (2.4) are satisfied. In addition we have

$$\gamma_{ij} k^j = \frac{1}{2}\gamma k_i, \tag{2.9}$$

and

$$R_{ijkm} = \frac{1}{2}\left(\ddot{\gamma}_{im} k_j k_k + \ddot{\gamma}_{jk} k_i k_m - \ddot{\gamma}_{ik} k_j k_m - \ddot{\gamma}_{jm} k_i k_k\right), \tag{2.10}$$

from which we deduce that

$$R_{ijkm} k^m = 0, \tag{2.11}$$

which indicates that the linearized Riemann tensor is type N (radiative type) in the Petrov classification with k^i as degenerate principal null direction. The null hypersurfaces $u = $ constant in Minkowskian space–time are generated by the null geodesic integral curves of the vector field k^i and are the histories of planes parallel to the xy-plane travelling with the speed of light in the positive z-direction. Thus $u = $ constant are the histories of the wavefronts of the plane waves. Now choose $\xi^i = \xi^i(u)$ so that the wave equation (2.6) is satisfied and the gauge transformation (2.7) reads

$$\bar{\gamma}_{ij} = \gamma_{ij} - \dot{\xi}_i k_j - \dot{\xi}_j k_i, \tag{2.12}$$

from which we obtain

$$\bar{\gamma}_{ij}^* = \gamma_{ij}^* - \dot{\xi}_i k_j - \dot{\xi}_j k_i + \eta_{ij}\dot{\xi}_k k^k. \tag{2.13}$$

12 Plane gravitational waves

We note that since $\gamma_{i3}^* + \gamma_{i4}^* = \gamma_{ij}^* k^j = 0$ we have $\bar{\gamma}_{i3}^* + \bar{\gamma}_{i4}^* = \bar{\gamma}_{ij}^* k^j = 0$. In view of (2.13) we see that we can choose $\xi_1(u)$ such that $\dot{\xi}_1 = \gamma_{13}^* = -\gamma_{14}^*$ in order to have $\bar{\gamma}_{13}^* = 0 = \bar{\gamma}_{14}^*$. We can also choose $\xi_2(u)$ such that $\dot{\xi}_2 = \gamma_{23}^* = -\gamma_{24}^*$ to have $\bar{\gamma}_{23}^* = 0 = \bar{\gamma}_{24}^*$. Finally it is convenient to choose $\xi_3(u)$ and $\xi_4(u)$ so that $\dot{\xi}_3 - \dot{\xi}_4 = \gamma_{33}^* = -\gamma_{34}^*$ and $\dot{\xi}_3 + \dot{\xi}_4 = -(\gamma_{11}^* + \gamma_{22}^*)/2$. The latter choices for ξ_3 and ξ_4 will ensure that $\bar{\gamma}_{ij}^*$ is trace-free and so $\bar{\gamma}_{ij}^* = \bar{\gamma}_{ij}$. Now, dropping the bars, we can say that without loss of generality $\gamma_{ij}(u)$ for plane waves travelling in the positive z-direction has the property that all components can be taken to vanish except for $\gamma_{11} = -\gamma_{22}$ and γ_{12}. The reader can verify that a further gauge transformation results in

$$\hat{\gamma}_{ij} = \gamma_{ij} - \lambda_{i,j} - \lambda_{j,i} = H\, k_i k_j, \tag{2.14}$$

and

$$H = \frac{1}{2}\ddot{\gamma}_{11}(x^2 - y^2) + \ddot{\gamma}_{12}\, x\, y, \tag{2.15}$$

with

$$\lambda_1 = \frac{1}{2}\{\gamma_{11}\, x + \gamma_{12}\, y\}, \tag{2.16}$$

$$\lambda_2 = \frac{1}{2}\{\gamma_{12}\, x - \gamma_{11}\, y\}, \tag{2.17}$$

$$\lambda_3 = -\frac{1}{4}\{\dot{\gamma}_{11}(x^2 - y^2) + 2\dot{\gamma}_{12}\, x\, y\}, \tag{2.18}$$

$$\lambda_4 = \frac{1}{4}\{\dot{\gamma}_{11}(x^2 - y^2) + 2\dot{\gamma}_{12}\, x\, y\}. \tag{2.19}$$

Clearly λ_i satisfies the wave equation (2.6). All components of λ_i are harmonic functions of x and y. Now the line-element of the space–time with metric tensor (2.1) can be written (dropping the hat in (2.14))

$$ds^2 = dx^2 + dy^2 + dz^2 - dt^2 + H\, du^2, \tag{2.20}$$

where $du = k_i\, dx^i$. With $u = z - t$ and $-2v = z + t$ this takes the slightly simpler form

$$ds^2 = dx^2 + dy^2 - 2\, du\, dv + H\, du^2. \tag{2.21}$$

Remarkably the metric tensor given by this line-element is an *exact* solution of Einstein's vacuum field equations and the space–time is a model of the gravitational field due to a train of plane gravitational waves of arbitrary profile. The fact that there are *two* arbitrary functions of u in H given by (2.15) demonstrates that the waves have two degrees of freedom of polarization just like plane electromagnetic waves.

We shall next transform the line-element (2.21) into a form in which the metric tensor components are independent of x and y. This is known as the Rosen (1937) form and, although the transformation can be given for the general case of $\ddot{\gamma}_{11} \neq 0$ and $\ddot{\gamma}_{12} \neq 0$ [see, for example, Futamase and Hogan (1993)] we will consider here the special case of $\ddot{\gamma}_{12} = 0$. The coordinate transformation is $(x, y, u, v) \to (x', y', u', v')$ given by

$$x = F(u')\, x', \tag{2.22}$$

$$x = G(u')\, y', \tag{2.23}$$
$$u = u', \tag{2.24}$$
$$v = v' + \frac{1}{2} F \dot{F} x'^2 + \frac{1}{2} G \dot{G} y'^2, \tag{2.25}$$

with the dot indicating differentiation with respect to u', and $F(u'), G(u')$ chosen to satisfy

$$\ddot{F} - \frac{1}{2}\ddot{\gamma}_{11}\, F = 0 \quad \text{and} \quad \ddot{G} + \frac{1}{2}\ddot{\gamma}_{11}\, G = 0. \tag{2.26}$$

The resulting Rosen form of the line-element (2.20) is

$$ds^2 = F^2 dx'^2 + G^2 dy'^2 - 2\, du'\, dv'. \tag{2.27}$$

From now on we shall drop the primes on the coordinates in (2.26) and (2.27). Our freedom to choose the function $\frac{1}{2}\ddot{\gamma}_{11}$ corresponds to our freedom to choose the profile of the plane gravitational waves. For a single plane *impulsive* wave we should choose $\frac{1}{2}\ddot{\gamma}_{11} = \delta(u)$, where $\delta(u)$ is the Dirac delta function which is singular on the null hypersurface $u = 0$. With this choice in (2.26) we find that

$$F = 1 + u\,\vartheta(u) \quad \text{and} \quad G = 1 - u\,\vartheta(u), \tag{2.28}$$

where $\vartheta(u)$ is the Heaviside step function which is equal to unity for $u > 0$ and vanishes for $u < 0$. Now the line-element reads

$$ds^2 = (1 + u\,\vartheta(u))^2 dx^2 + (1 - u\,\vartheta(u))^2 dy^2 - 2\, du\, dv. \tag{2.29}$$

The metric given via this line-element is a solution of the vacuum field equations $R_{ij} = 0$, where R_{ij} are the components of the Ricci tensor. These field equations hold everywhere, in particular on $u = 0$. The non-identically vanishing components of the Riemann tensor are proportional to $\delta(u)$ (thus $R_{ijkl} \propto \delta(u)$). Hence we see that if $u < 0$ or $u > 0$ then (2.29) is the line-element of Minkowskian space–time (vanishing Riemann tensor) and the Riemann tensor is singular on the null hypersurface history $u = 0$ of the plane impulsive gravitational wave. In the space–time with line-element (2.29) there are two families of intersecting null hypersurfaces, $u = $ constant and $v = $ constant. Thus a plane, homogeneous, impulsive gravitational wave propagating in the opposite direction to that with history $u = 0$ above, with history in space–time $v = 0$, is described by a space–time with line-element

$$ds^2 = (1 + v\,\vartheta(v))^2 dx^2 + (1 - v\,\vartheta(v))^2 dy^2 - 2\, du\, dv. \tag{2.30}$$

Following the collision of two such waves we have $u > 0$ and $v > 0$ and the gravitational field is described by a vacuum space–time with line-element of the form (Khan and Penrose 1971, Szekeres 1970, 1972)

$$ds^2 = e^{-U+V} dx^2 + e^{-U-V} dy^2 - 2\, e^{-M} du\, dv, \tag{2.31}$$

where U, V, M are functions of u and v. A simple but important observation about this line-element is that under the coordinate transformation $u \to \bar{u}(u)$ and $v \to \bar{v}(v)$

14 Plane gravitational waves

the form of the line-element is unchanged but the function M is transformed to $\bar{M}$ with

$$e^{-\bar{M}} = e^{-M}\frac{du}{d\bar{u}}\frac{dv}{d\bar{v}}. \tag{2.32}$$

The region of space–time $u < 0$ has line-element (2.30) and the region of space–time $v < 0$ has line-element (2.29). The space–time model of the vacuum gravitational field after the collision has line-element of the form (2.31) with $u > 0, v > 0$ and with the functions U, V, M satisfying the following boundary conditions:

On $v = 0, u > 0$:

$$U = -\log(1 - u^2), \quad V = \log\left(\frac{1+u}{1-u}\right), \quad M = 0; \tag{2.33}$$

On $u = 0, v > 0$:

$$U = -\log(1 - v^2), \quad V = \log\left(\frac{1+v}{1-v}\right), \quad M = 0. \tag{2.34}$$

Clearly these conditions ensure continuity of the metric tensor components on the boundaries $v = 0, u > 0$ and $u = 0, v > 0$ of the post-collision region of the space–time. With subscripts denoting partial derivatives, the vacuum field equations to be satisfied by the functions U, V, M in (2.31) read:

$$U_{uv} = U_u\, U_v, \tag{2.35}$$

$$2\, V_{uv} = U_u\, V_v + U_v\, V_u, \tag{2.36}$$

$$2\, U_{uu} = U_u^2 + V_u^2 - 2\, U_u\, M_u, \tag{2.37}$$

$$2\, U_{vv} = U_v^2 + V_v^2 - 2\, U_v\, M_v, \tag{2.38}$$

$$2\, M_{uv} = V_u\, V_v - U_u\, U_v. \tag{2.39}$$

The first of these equations can be written $(e^{-U})_{uv} = 0$ and this is easy to integrate to

$$e^{-U} = f(u) + g(u), \tag{2.40}$$

and so to satisfy (2.33) and (2.34) we have

$$e^{-U} = 1 - u^2 - v^2. \tag{2.41}$$

Now V is calculated from (2.36) while M is given by (2.37) and (2.38) with (2.39) the integrability condition (or the consistency condition) for (2.37) and (2.38). To solve (2.36) we write it in a way that suggests a simplifying assumption (see Barrabès and Hogan (2003b)), namely,

$$2\frac{\partial^2}{\partial u\, \partial v}\log\left(\frac{V_v}{V_u}\right) = \left(U_v\,\frac{V_u}{V_v}\right)_v - \left(U_u\,\frac{V_v}{V_u}\right)_u. \tag{2.42}$$

The simplifying assumption that this equation suggests is to try the separation of variables

$$\frac{V_v}{V_u} = \frac{A(u)}{B(v)}. \tag{2.43}$$

To determine the functions $A(u)$ and $B(v)$ we only require the boundary values of V_u and V_v which we can calculate from (2.36). For example (2.36) evaluated at $v = 0$ and using (2.33) and (2.41) yields the differential equation

$$\frac{d}{du}(V_v)_{v=0} = \frac{u}{1-u^2}(V_v)_{v=0}, \qquad (2.44)$$

and so

$$(V_v)_{v=0} = \frac{a_0}{\sqrt{1-u^2}}, \qquad (2.45)$$

with a_0 a constant of integration. But (2.34) gives

$$(V_v)_{u=0} = \frac{2}{1-v^2}, \qquad (2.46)$$

and thus for (2.45) and (2.46) to agree when $v = 0$ *and* when $u = 0$ we must have $a_0 = 2$. Hence

$$(V_v)_{v=0} = \frac{2}{\sqrt{1-u^2}}. \qquad (2.47)$$

Similarly we have

$$(V_u)_{v=0} = \frac{2}{1-u^2} \quad \text{and} \quad (V_u)_{u=0} = \frac{2}{\sqrt{1-v^2}}. \qquad (2.48)$$

Now evaluating (2.43) at $u = 0$ and at $v = 0$ will determine the functions $A(u)$ and $B(v)$. We easily find that

$$\frac{V_v}{V_u} = \frac{\sqrt{1-u^2}}{\sqrt{1-v^2}}. \qquad (2.49)$$

Introducing the coordinates $\bar{u}, \bar{v}$ via

$$\bar{u} = \sin^{-1} u \quad \text{and} \quad \bar{v} = \sin^{-1} v, \qquad (2.50)$$

we see that (2.49) becomes the first-order wave equation

$$V_{\bar{u}} = V_{\bar{v}}, \qquad (2.51)$$

which immediately integrates to $V = V(\bar{u} + \bar{v})$. When $v = 0$ we have $\bar{v} = 0$ and, by (2.33) and (2.50), we have

$$V = \log\left(\frac{1+\sin \bar{u}}{1-\sin \bar{u}}\right) \quad \text{when} \quad \bar{v} = 0. \qquad (2.52)$$

Hence for $\bar{u} > 0$ and $\bar{v} > 0$ we arrive at

$$V(\bar{u}, \bar{v}) = \log\left(\frac{1+\sin(\bar{u}+\bar{v})}{1-\sin(\bar{u}+\bar{v})}\right) = \log\left\{\left(\frac{\cos \bar{u} + \sin \bar{v}}{\cos \bar{u} - \sin \bar{v}}\right)\left(\frac{\cos \bar{v} + \sin \bar{u}}{\cos \bar{v} - \sin \bar{u}}\right)\right\}. \qquad (2.53)$$

The equality of the arguments of the logarithms here is a nice trigonometric identity to establish. In terms of the barred coordinates we see that (2.41) reads

$$e^{-U} = \cos(\bar{u} - \bar{v})\cos(\bar{u} + \bar{v}), \qquad (2.54)$$

Plane gravitational waves

from which it follows that U satisfies the second-order wave equation

$$U_{\bar u \bar u} = U_{\bar v \bar v}. \tag{2.55}$$

The reader can readily check that this equation is equivalent to the vanishing of the right-hand side of (2.42). To calculate M we first calculate $\bar M$, using the field equations written in terms of the barred coordinates and $\bar M$, and then obtain M from (2.32). It is convenient to define

$$\bar Q = \bar M + \frac{1}{2} U, \tag{2.56}$$

and then (2.37) and (2.38) in the barred variables reduce to

$$\bar Q_{\bar u} = \bar Q_{\bar v} = 2 \tan(\bar u + \bar v). \tag{2.57}$$

We note at this point that (2.39) in the barred variables reads

$$\bar Q_{\bar u \bar v} = \frac{1}{2} V_{\bar u} V_{\bar v}, \tag{2.58}$$

and this is now satisfied on account of (2.57) and V given by (2.53). To solve (2.57) we need the boundary conditions: When $\bar u = 0$ we must have $\bar Q = -2 \log \cos \bar v$ and when $\bar v = 0$ we must have $\bar Q = -2 \log \cos \bar u$, which follow easily from the boundary conditions above expressed in terms of the barred coordinates. We thus obtain from (2.57) the solution

$$\bar Q = -2 \log \cos(\bar u + \bar v). \tag{2.59}$$

With this and (2.47) and (2.49) we have

$$e^{\bar M} = \left(\frac{\cos(\bar u - \bar v)}{\cos^3(\bar u + \bar v)} \right)^{1/2}. \tag{2.60}$$

Now (2.32) with (2.50) yields

$$e^{-M} = e^{-\frac{3}{2}U} \{\cos^2(\bar u - \bar v) \cos \bar u \cos \bar v\}^{-1}. \tag{2.61}$$

The functions U, V, M in (2.31) for $u > 0, v > 0$ are given by (2.54), (2.53), and (2.61), respectively. Writing them in terms of the coordinates u, v we have U given by (2.41), V is now

$$e^V = \left(\frac{\sqrt{1-u^2}+v}{\sqrt{1-u^2}-v} \right) \left(\frac{\sqrt{1-v^2}+u}{\sqrt{1-v^2}-u} \right), \tag{2.62}$$

and M is given by

$$e^{-M} = \frac{(1-u^2-v^2)^{3/2}}{\{\sqrt{1-u^2}\sqrt{1-v^2}+uv\}^2 \sqrt{1-u^2}\sqrt{1-v^2}}. \tag{2.63}$$

When (2.41), (2.62), and (2.63) are substituted into the line-element (2.31) we arrive at the Khan–Penrose (1971) solution of Einstein's vacuum field equations. This expression for the line-element is valid for $u > 0, v > 0$. To obtain an expression valid for all u, v we simply replace u and v in (2.41), (2.62), and (2.63) by $u_+ = u\,\vartheta(u)$ and $v_+ = v\,\vartheta(v)$, respectively. The reader can verify that the Khan–Penrose solution for $u > 0, v > 0$ has a singularity in the curvature tensor at $e^{-U} = 1 - u^2 - v^2 = 0$ and so the solution is valid in the region $u > 0, v > 0$ only up to the quadrant of the circle $u^2 + v^2 = 1$. Further properties of this solution can be found in Griffiths (1991).

2.2 Electromagnetic shock waves

Henceforth in this chapter we shall only consider line-elements of the simple form (with $M = 0$)

$$ds^2 = e^{-U+V} dx^2 + e^{-U-V} dy^2 - 2\, du\, dv, \tag{2.64}$$

where U and V are in general functions of u and v. We can write (2.64) as

$$ds^2 = (\vartheta^1)^2 + (\vartheta^2)^2 - 2\,\vartheta^3 \vartheta^4 = g_{ab} \vartheta^a \vartheta^b, \tag{2.65}$$

with the basis 1-forms defined by

$$\vartheta^1 = e^{(-U+V)/2} dx = \vartheta_1, \tag{2.66}$$
$$\vartheta^2 = e^{-(U+V)/2} dy = \vartheta_2, \tag{2.67}$$
$$\vartheta^3 = du = -\vartheta_4, \tag{2.68}$$
$$\vartheta^4 = dv = -\vartheta_3. \tag{2.69}$$

These 1-forms define a half-null tetrad. The constants g_{ab} are the components of the metric tensor on this tetrad and tetrad indices are lowered [as in the second equalities in (2.66)–(2.69)] and raised using g_{ab} and its inverse $g^{ab} = g_{ab}$, respectively. With subscripts on U, V denoting partial differentiation the non-vanishing components of the Riemann curvature tensor on the half-null tetrad are

$$R_{1212} = \frac{1}{2}(U_u U_v - V_u V_v), \tag{2.70}$$

$$R_{1313} = \frac{1}{2}(U_{uu} - V_{uu}) - \frac{1}{4}(U_u - V_u)^2, \tag{2.71}$$

$$R_{1314} = \frac{1}{2}(U_{uv} - V_{uv}) - \frac{1}{4}(U_u - V_u)(U_v - V_v), \tag{2.72}$$

$$R_{2323} = \frac{1}{2}(U_{uu} + V_{uu}) - \frac{1}{4}(U_u + V_u)^2, \tag{2.73}$$

$$R_{2324} = \frac{1}{2}(U_{uv} + V_{uv}) - \frac{1}{4}(U_u + V_u)(U_v + V_v), \tag{2.74}$$

$$R_{1414} = \frac{1}{2}(U_{vv} - V_{vv}) - \frac{1}{4}(U_v - V_v)^2, \tag{2.75}$$

$$R_{2424} = \frac{1}{2}(U_{vv} + V_{vv}) - \frac{1}{4}(U_v + V_v)^2. \tag{2.76}$$

The non-identically vanishing components of the Ricci tensor on the half-null tetrad, $R_{ab} = g^{cd} R_{acbd}$, are given by

$$R_{11} = -U_{uv} + U_u U_v + V_{uv} - \frac{1}{2}(U_u V_v + U_v V_u), \tag{2.77}$$

18 Plane gravitational waves

$$R_{22} = -U_{uv} + U_u U_v - V_{uv} + \frac{1}{2}(U_u V_v + U_v V_u), \tag{2.78}$$

$$R_{33} = U_{uu} - \frac{1}{2}(U_u^2 + V_u^2), \tag{2.79}$$

$$R_{34} = U_{uv} - \frac{1}{2}(U_u U_v + V_u V_v), \tag{2.80}$$

$$R_{44} = U_{vv} - \frac{1}{2}(U_v^2 + V_v^2). \tag{2.81}$$

The Weyl conformal curvature tensor components C_{abcd} on the tetrad are related to the components R_{abcd} of the Riemann curvature tensor on the tetrad, the components R_{ab} of the Ricci tensor on the tetrad, and the Ricci scalar $R = g^{ab} R_{ab}$ by the formula

$$C_{abcd} = R_{abcd} + \frac{1}{2}(g_{ad} R_{bc} + g_{bc} R_{ad} - g_{ac} R_{bd} - g_{bd} R_{ac})$$
$$+ \frac{1}{6} R (g_{ac} g_{bd} - g_{ad} g_{bc}). \tag{2.82}$$

We note that the Newman–Penrose (1962) components Ψ_A for $A = 0, 1, 2, 3, 4$ of the Weyl conformal curvature tensor are related to the tetrad components of the Riemann and Ricci tensors by

$$\Psi_0 = R_{1313} - \frac{1}{2} R_{33} + i R_{1323}, \tag{2.83}$$

$$\Psi_1 = \frac{1}{\sqrt{2}}(R_{3431} + i R_{3432}) - \frac{1}{2\sqrt{2}}(R_{31} + i R_{32}), \tag{2.84}$$

$$\Psi_2 = \frac{1}{2}\left(R_{3434} + i R_{3412} - R_{34} + \frac{1}{6} R\right), \tag{2.85}$$

$$\Psi_3 = \frac{1}{\sqrt{2}}\left(R_{3414} - i R_{3424} + \frac{1}{2} R_{41} + \frac{1}{2} i R_{42}\right), \tag{2.86}$$

$$\Psi_4 = R_{1414} - \frac{1}{2} R_{44} - i R_{1424}, \tag{2.87}$$

with $R = R_{11} + R_{22} - 2 R_{34}$.

A quick perusal of the passage from (2.21) to (2.27) will reveal that a similar coordinate transformation applied to (2.21) when

$$H = A(u)(x^2 + y^2), \tag{2.88}$$

will also lead, after dropping the primes, to a homogeneous metric tensor (having components independent of x and y) given by the line-element

$$ds^2 = F^2 dx^2 + G^2 dy^2 - 2 du\, dv, \tag{2.89}$$

where $F(u)$ and $G(u)$ satisfy

$$\ddot{F} - A F = 0 \quad \text{and} \quad \ddot{G} - A G = 0. \tag{2.90}$$

A simple solution corresponding to the choice $A(u) = -a^2 \vartheta(u)$, where $\vartheta(u)$ is the Heaviside step function used above and a is a constant, is given by

$$F(u) = G(u) = \cos a\, u_+, \tag{2.91}$$

where $u_+ = u\,\vartheta(u)$. In differentiating this function we note that $d\vartheta(u)/du = \delta(u)$ and also $f(u)\,\delta(u) = f(0)\,\delta(u)$ and $\vartheta^2(u) = \vartheta(u)$. The resulting line-element

$$ds^2 = \cos^2 au_+\,(dx^2 + dy^2) - 2\,du\,dv, \tag{2.92}$$

is interesting from a physical point of view. It fits into the expression (2.64) and therefore the calculation of the Ricci tensor is easily carried out using the formulae (2.77)–(2.81) to obtain

$$R_{ab} = 2a^2\vartheta(u)\delta_a^3\,\delta_b^3. \tag{2.93}$$

These are the vacuum Einstein–Maxwell field equations

$$R_{ab} = 2\,E_{ab}, \tag{2.94}$$

with the electromagnetic energy tensor

$$E_{ab} = F_{ac}F_b{}^c - \frac{1}{4}\,g_{ab}\,F_{cd}\,F^{cd}, \tag{2.95}$$

derived from the Maxwell 2-form

$$F = \frac{1}{2}\,F_{ab}\,\vartheta^a \wedge \vartheta^b = a\,\vartheta(u)\,\vartheta^1 \wedge \vartheta^3. \tag{2.96}$$

It is simple to check that this 2-form is a solution of Maxwell's vacuum field equations $dF = 0 = d^*F$, where d is the exterior derivative and the star indicates the Hodge dual. This Maxwell field is type N (the radiative type) in the Petrov classification of 2-forms and describes electromagnetic radiation with propagation direction in space–time given by the vector field $\partial/\partial v$. The profile of the wave is the step function and so we have here an electromagnetic shock wave.

The head-on collision of two electromagnetic shock waves is described by the Bell–Szekeres (1974) solution of the vacuum Einstein–Maxwell field equations with line-element

$$ds^2 = \cos^2(au_+ + bv_+)\,dx^2 + \cos^2(au_+ - bv_+)\,dy^2 - 2\,du\,dv, \tag{2.97}$$

where b is a constant and $v_+ = v\,\vartheta(v)$. When $v < 0$ this coincides with (2.92) and when $u < 0$ it describes the second incoming electromagnetic shock wave with propagation direction in space–time given by the vector field $\partial/\partial u$. The region of space–time corresponding to $u < 0$ *and* $v < 0$ is Minkowskian while the region $u > 0, v > 0$ corresponds to the post-collision and has line-element given by (2.97) with $u_+ = u$ and $v_+ = v$. Calculation of the tetrad components of the Ricci tensor using the formulae above yields the only non-identically vanishing components to be

$$R_{11} = -R_{22} = -2a\,b\,\vartheta(u)\,\vartheta(v)\,,\quad R_{33} = 2\,a^2\vartheta(u)\,,\quad R_{44} = 2\,b^2\vartheta(v). \tag{2.98}$$

These are the vacuum Einstein–Maxwell field equations (2.94) with the electromagnetic energy tensor calculated from the Maxwell 2-form

$$F = a\,\vartheta(u)\,\vartheta^1 \wedge \vartheta^3 + b\,\vartheta(v)\,\vartheta^1 \wedge \vartheta^4. \tag{2.99}$$

This 2-form is easily seen to satisfy Maxwell's vacuum field equations. We see in (2.99) the incoming electromagnetic shock waves in the regions $u < 0$ and $v < 0$ and that the

20 Plane gravitational waves

electromagnetic field in the post-collision region $u > 0, v > 0$ is a simple superposition. The Newman–Penrose components of the Weyl conformal curvature tensor vanish except for

$$\Psi_0 = a\,\delta(u)\,\tan bv_+ \quad \text{and} \quad \Psi_4 = b\,\delta(v)\,\tan au_+. \tag{2.100}$$

This shows that before and after the collision the space–time is conformally flat and following the collision impulsive gravitational waves, one with history $u = 0, v > 0$, described by Ψ_0, and another with history $v = 0, u > 0$, described by Ψ_4, are created. These products of the collision of the electromagnetic shock waves constitute a redistribution, following the collision, of the energy in the incoming shock waves. Further properties of the Bell–Szekeres solution can be found in Griffiths (1991).

When $u > 0, v > 0$ we make the following coordinate transformations on the line-element (2.97): if $ab > 0$ define coordinates ξ, η by $\sqrt{2ab}\,\xi = au + bv$ and $\sqrt{2ab}\,\eta = au - bv$ then (2.97) when $u > 0, v > 0$ reads

$$ds^2 = g'_{AB}\,dx^A\,dx^B + g''_{AB}\,dy^A\,dy^B, \tag{2.101}$$

with

$$g'_{AB}\,dx^A\,dx^B = -d\xi^2 + \cos^2(\sqrt{2ab}\,\xi)\,dx^2, \tag{2.102}$$

$$g''_{AB}\,dy^A\,dy^B = d\eta^2 + \cos^2(\sqrt{2ab}\,\eta)\,dy^2, \tag{2.103}$$

and with capital letters taking values 1, 2 and $x^A = (\xi, x)$, $y^A = (\eta, y)$. Thus (2.101) indicates that the Bell–Szekeres manifold (with line-element given by (2.97) with $u > 0, v > 0$) is the Cartesian product of two 2-dimensional manifolds. Calculation of the Riemann curvature tensor for (2.102) and for (2.103) reveals the forms

$$R'_{ABCD} = 2ab\,(g'_{AD}\,g'_{BC} - g'_{AC}\,g'_{BD}), \tag{2.104}$$

$$R''_{ABCD} = -2ab\,(g''_{AD}\,g''_{BC} - g''_{AC}\,g''_{BD}), \tag{2.105}$$

respectively, indicating that the 2-dimensional manifolds have constant curvature of opposite signs. On the other hand, if $ab < 0$ then defining coordinates ξ, η by $\sqrt{-2ab}\,\xi = au + bv$ and $\sqrt{2ab}\,\eta = au - bv$ results again in (2.101) but now with

$$g'_{AB}\,dx^A\,dx^B = d\xi^2 + \cos^2(\sqrt{-2ab}\,\xi)\,dx^2, \tag{2.106}$$

$$g''_{AB}\,dy^A\,dy^B = -d\eta^2 + \cos^2(\sqrt{-2ab}\,\eta)\,dy^2. \tag{2.107}$$

Calculation of the Riemann curvature tensor components in each case again results in (2.104) and (2.105) so that again the two 2-dimensional manifolds each have constant curvature of equal and opposite sign but with the signs of the curvatures reversed compared to those of (2.102) and (2.103) because now $ab < 0$. The space–time with line-element (2.101), with g'_{AB} and g''_{AB} given either by (2.102) and (2.103) or by (2.106) and (2.107), is a Bertotti–Robinson (Bertotti 1959, Robinson 1959) space–time. This property of the Bell–Szekeres space–time is well known (Stephani et al. 2003, p. 399). It is the prime motivation for the discussion in the next section.

2.3 Gravitational shock waves

The Bertotti–Robinson space–time described above is a homogeneous solution of the vacuum Einstein–Maxwell field equations. The so-called Nariai–Bertotti (Nariai 1999, Bertotti 1959) space–time is a homogeneous solution of Einstein's vacuum field equations with a cosmological constant Λ. This latter space–time manifold is also the Cartesian product of two 2-dimensional manifolds with line-element of the form (2.101) but for which the two 2-dimensional manifolds have the *same* constant curvature (rather than having constant curvatures of opposite signs). It has recently been demonstrated (Barrabès and Hogan, 2011) that the Nariai–Bertotti space–time coincides with the post-collision space–time following the head-on collision of two plane, homogeneous, *gravitational* shock waves. This demonstration begins with the case $\Lambda < 0$. A convenient representation of the Nariai–Bertotti line-element is given in this case by (2.101) with

$$g'_{AB} dx^A dx^B = -d\xi^2 + \cos^2(\sqrt{-\Lambda}\,\xi)\, dx^2, \tag{2.108}$$

$$g''_{AB} dy^A dy^B = d\eta^2 + \cosh^2(\sqrt{-\Lambda}\,\eta)\, dy^2, \tag{2.109}$$

with $x^A = (\xi, x)$, $y^A = (\eta, y)$. The Riemann curvature tensors for these 2-dimensional manifolds are given by

$$R'_{ABCD} = -\Lambda\,(g'_{AD}\, g'_{BC} - g'_{AC}\, g'_{BD}), \tag{2.110}$$

$$R''_{ABCD} = -\Lambda\,(g''_{AD}\, g''_{BC} - g''_{AC}\, g''_{BD}), \tag{2.111}$$

indicating that the manifolds with line-elements (2.108) and (2.109) have *equal* constant curvatures. Now put $\Lambda = -2g_0 g_1$, where g_0, g_1 are real constants and define new coordinates u, v in place of ξ, η by

$$\sqrt{-\Lambda}\,\xi = g_0 u + g_1 v, \tag{2.112}$$

$$\sqrt{-\Lambda}\,\eta = g_0 u - g_1 v. \tag{2.113}$$

Making these transformations in (2.108) and (2.109) and then substituting the results into (2.101) results in

$$ds^2 = \cos^2(g_0 u + g_1 v)\, dx^2 + \cosh^2(g_0 u - g_1 v)\, dy^2 - 2\, du\, dv. \tag{2.114}$$

For the case $\Lambda > 0$ the line-elements (2.108) and (2.109) are replaced by

$$g'_{AB} dx^A dx^B = d\xi^2 + \cos^2(\sqrt{\Lambda}\,\xi)\, dx^2, \tag{2.115}$$

$$g''_{AB} dy^A dy^B = -d\eta^2 + \cosh^2(\sqrt{\Lambda}\,\eta)\, dy^2. \tag{2.116}$$

Now (2.110) and (2.111) take the same form so that in this case the two 2-dimensional manifolds have equal constant curvatures but of opposite sign to the equal constant curvatures in the case of $\Lambda < 0$. In this case the transformations (2.112) and (2.113) are replaced by

$$\sqrt{\Lambda}\,\xi = g_0 u + g_1 v, \tag{2.117}$$

$$\sqrt{\Lambda}\,\eta = g_0 u - g_1 v. \tag{2.118}$$

22 Plane gravitational waves

Making these transformations in (2.115) and (2.116) and then substituting the results into (2.101) results again in (2.114).

Now we wish to consider (2.114) to be the result of a collision and to effect this we replace u by $u_+ = u\,\vartheta(u)$ and v by $v_+ = v\,\vartheta(v)$ in the metric tensor components in (2.114). Hence we consider the line-element

$$ds^2 = \cos^2(g_0 u_+ + g_1 v_+)\,dx^2 + \cosh^2(g_0 u_+ - g_1 v_+)\,dy^2 - 2\,du\,dv,$$
$$= (\vartheta^1)^2 + (\vartheta^2)^2 - 2\,\vartheta^3\vartheta^4,$$
$$= g_{ab}\vartheta^a\vartheta^b, \tag{2.119}$$

with the constants g_{ab} the components of the metric tensor on the half-null tetrad defined via the basis 1-forms, with the 1-forms $\{\vartheta^a\}$, for $a = 1, 2, 3, 4$ defined by

$$\vartheta^1 = \cos(g_0 u_+ + g_1 v_+)\,dx\ ,\ \vartheta^2 = \cosh(g_0 u_+ - g_1 v_+)\,dy\ ,\ \vartheta^3 = du\ ,\ \vartheta^4 = dv. \tag{2.120}$$

This fits the pattern of (2.66)–(2.69) with

$$e^{-U} = \cos(g_0 u_+ + g_1 v_+)\cosh(g_0 u_+ - g_1 v_+), \tag{2.121}$$

$$e^V = \frac{\cos(g_0 u_+ + g_1 v_+)}{\cosh(g_0 u_+ - g_1 v_+)}. \tag{2.122}$$

We now use (2.77)–(2.81) to calculate the tetrad components of the Ricci tensor and (2.70)–(2.76) followed by (2.83)–(2.87) to evaluate the Newman–Penrose components of the Weyl conformal curvature tensor. As a guide to the reader in carrying out these calculations we give the following partial derivatives as examples:

$$U_u = g_0 \vartheta(u)\{\tan(g_0 u_+ + g_1 v_+) - \tanh(g_0 u_+ - g_1 v_+)\}, \tag{2.123}$$

$$U_v = g_1 \vartheta(v)\{\tan(g_0 u_+ + g_1 v_+) + \tanh(g_0 u_+ - g_1 v_+)\}, \tag{2.124}$$

and

$$U_{uv} = g_0 g_1 \vartheta(u)\vartheta(v)\{\sec^2(g_0 u_+ + g_1 v_+) + \text{sech}^2(g_0 u_+ - g_1 v_+)\}, \tag{2.125}$$

$$U_{uu} = g_0 \delta(u)\{\tan g_1 v_+ + \tanh g_1 v_+\} + g_0^2 \vartheta(u)\{\sec^2(g_0 u_+ + g_1 v_+) - \text{sech}^2(g_0 u_+ - g_1 v_+)\}, \tag{2.126}$$

$$U_{vv} = g_1 \delta(v)\{\tan g_0 u_+ + \tanh g_0 u_+\} + g_1^2 \vartheta(v)\{\sec^2(g_0 u_+ + g_1 v_+) - \text{sech}^2(g_0 u_+ - g_1 v_+)\}. \tag{2.127}$$

The Ricci tensor components on the half-null tetrad are given by

$$R_{ab} = \Lambda\,\vartheta(u)\vartheta(v)\,g_{ab} - g_0\delta(u)\{\tan g_1 v_+ + \tanh g_1 v_+\}\delta^3_a\delta^3_b$$
$$-g_1\delta(v)\{\tan g_0 u_+ + \tanh g_0 u_+\}\delta^4_a\delta^4_b, \tag{2.128}$$

with

$$\Lambda = -2 g_0 g_1. \tag{2.129}$$

Thus the regions of the space–time with line-element (2.119) for which $u < 0$ and for which $v < 0$ are vacuum space–times (the pre-collision regions). The region for which $u > 0$ *and* $v > 0$ (the post-collision space–time) is a solution of Einstein's vacuum field equations with a cosmological constant:

$$R_{ab} = \Lambda\, g_{ab}. \tag{2.130}$$

The delta function terms in (2.128) describe light-like shells of matter [such as bursts of neutrinos for example, see Barrabès and Hogan (2003b)] with histories in space–time given by $u = 0, v > 0$ (the term with coefficient g_0) and by $v = 0, u > 0$ (the term with coefficient g_1). The Newman–Penrose components of the Weyl conformal curvature tensor are given by

$$\Psi_0 = \frac{1}{2} g_0\, \delta(u)\{\tan g_1 v_+ - \tanh g_1 v_+\} + g_0^2\, \vartheta(u), \tag{2.131}$$

$$\Psi_1 = 0, \tag{2.132}$$

$$\Psi_2 = \frac{1}{3} g_0\, g_1 \vartheta(u)\, \vartheta(v), \tag{2.133}$$

$$\Psi_3 = 0, \tag{2.134}$$

$$\Psi_4 = \frac{1}{2} g_1\, \delta(v)\{\tan g_0 u_+ - \tanh g_0 u_+\} + g_1^2\, \vartheta(v). \tag{2.135}$$

Here the delta function terms represent impulsive gravitational waves created after the collision. The wave in Ψ_0 has history in space–time the portion of a null hypersurface $u = 0, v > 0$ and the wave in Ψ_4 has history in space–time the portion of the null hypersurface $v = 0, u > 0$. Before the collision in the region $v < 0$ we see that the only non-vanishing component of the Weyl tensor is $\Psi_0 = g_0^2\, \vartheta(u)$. This is a vacuum region, as noted following (2.129) above, and the Weyl tensor therefore coincides with the Riemann tensor which is type N (radiative type) in the Petrov classification with $\partial/\partial v$ as degenerate principal null direction. The profile of the wave is proportional to $\vartheta(u)$ and so this is an incoming plane, homogeneous gravitational shock wave. Similarly in the region $u < 0$ we have another incoming plane, homogeneous gravitational shock wave with propagation direction $\partial/\partial u$ described by the step function $\Psi_4 = g_1^2\, \vartheta(v)$. The post-collision region is $u > 0, v > 0$ and there the non-vanishing Weyl tensor components are given by $\Psi_0 = g_0^2, \Psi_2 = g_0 g_1/3, \Psi_4 = g_1^2$ and this gravitational field is the Petrov type D Nariai–Bertotti homogeneous solution of Einstein's vacuum field equations with a cosmological constant.

We have seen in the previous section how the energy in the incoming electromagnetic shock waves is redistributed after the collision in such a way that two impulsive gravitational waves are created. The situation following the collision of the gravitational shock waves in this section is somewhat more dramatic. The redistribution of the energy in the incoming waves in this case takes the form of the creation of two impulsive gravitational waves, two light-like shells and a cosmological constant (in the post-collision region). The cosmological constant also represents a form of energy (so-called *dark energy*) since it can be viewed as describing a perfect fluid matter distribution in which the isotropic pressure p and proper density μ satisfy the equation of

24 Plane gravitational waves

state $p + \mu = 0$. It is possible to combine the electromagnetic and gravitational shock waves into a single light-like signal. A head-on collision of such signals has been studied by Barrabès and Hogan (2011).

2.4 High-frequency gravity waves

For modelling astrophysical processes two of the most useful families of gravitational waves are bursts of gravitational radiation, perhaps accompanied by matter travelling with the speed of light, such as neutrinos, and high-frequency gravitational waves. For the case of high-frequency waves the fundamental building blocks are monochromatic waves. We are concerned here with approximate solutions of Einstein's vacuum field equations in which the approximations are controlled by a small parameter λ (say) which plays the role of the wavelength of the radiation. The line-element for the approximate vacuum space–time model of the gravitational field of a train of homogeneous, monochromatic, plane gravitational waves can be put in the form (Burnett, 1989)

$$ds^2 = 2\, B_\lambda(u)^2 \left| d\zeta + \lambda\, \bar{W}(u) \sin \frac{u}{\lambda}\, d\bar{\zeta} \right|^2 - 2\, du\, dv. \tag{2.136}$$

Here $B_\lambda(u)$ is a real-valued function of the real coordinate u which also depends upon the real parameter $\lambda \geq 0$. $W(u)$ is an arbitrary complex-valued function of u (the bar, as always, denotes complex conjugation). The hypersurfaces $u =$ constant are null and are generated by the null geodesic integral curves of the vector field $\partial/\partial v$ with v real and an affine parameter along them. The null hypersurfaces are the histories of the plane wavefronts of the gravitational waves. For calculations it is natural to start with the null tetrad defined by the basis 1-forms

$$\omega^1 = B_\lambda(u) \left(d\zeta + \lambda\, \bar{W}(u) \sin \frac{u}{\lambda}\, d\bar{\zeta} \right), \tag{2.137}$$

$$\omega^2 = \bar{\omega}^1, \tag{2.138}$$

$$\omega^3 = du, \tag{2.139}$$

$$\omega^4 = dv. \tag{2.140}$$

The components R_{ab} of the Ricci tensor calculated on this null tetrad satisfy

$$R_{ab} = O(\lambda), \tag{2.141}$$

and in this sense the vacuum field equations are approximately satisfied for small λ, provided $B_\lambda(u)$ satisfies

$$\ddot{B}_\lambda + |W|^2 \sin^2 \frac{u}{\lambda}\, B_\lambda = 0, \tag{2.142}$$

with the dots denoting differentiation with respect to u. The Newman–Penrose components of the Riemann curvature tensor on the null tetrad satisfy

$$\Psi_0 = \lambda^{-1} W \sin \frac{u}{\lambda} + O(\lambda^0) \quad \text{and} \quad \Psi_A = O(\lambda) \quad \text{for} \quad A = 1, 2, 3, 4. \tag{2.143}$$

Thus for small λ the Riemann tensor is type N in the Petrov classification with $\partial/\partial v$ the degenerate principal null direction. In addition the profile of the waves has large amplitude and short wavelength (high frequency).

Let us suppose that the high-frequency waves exist ($W(u) \neq 0$) for a finite interval $u_1 \leq u \leq u_2$. Let u' be any value of u in this interval and let $\vartheta(u - u')$ be the Heaviside step function which is equal to unity for $u - u' > 0$ and which vanishes for $u - u' < 0$. We can use this function as a Greens' function for the differential equation (2.142). Multiplying (2.142) by the step function and integrating by parts results in the integral equation

$$\dot{B}_\lambda(u') = \dot{B}_\lambda(u_2) + \int_{u_1}^{u_2} \vartheta(u - u') \, |W(u)|^2 \, B_\lambda(u) \sin^2 \frac{u}{\lambda} \, du. \tag{2.144}$$

We assume that $B_\lambda(u)$ has a uniform $\lambda = 0$ limit on the interval $u_1 \leq u \leq u_2$. To take the limit of (2.144) we use the Riemann–Lebesgue theorem (Olmsted, 1959) which states that if a real-valued function $A(u)$ is integrable (and therefore could be a step function) on the interval $u_1 \leq u \leq u_2$ then

$$\lim_{\lambda \to 0} \int_{u_1}^{u_2} A(u) \cos \frac{u}{\lambda} \, du = 0. \tag{2.145}$$

Hence it follows that

$$\lim_{\lambda \to 0} \int_{u_1}^{u_2} A(u) \sin^2 \frac{u}{\lambda} \, du = \frac{1}{2} \int_{u_1}^{u_2} A(u) \, du. \tag{2.146}$$

Thus taking the limit $\lambda \to 0$ of the equation (2.144) results in the integral equation

$$\dot{B}_0(u') = \dot{B}_0(u_2) + \frac{1}{2} \int_{u_1}^{u_2} \vartheta(u - u') \, |W(u)|^2 \, B_0(u) \, du. \tag{2.147}$$

Differentiating this equation with respect to u', using $d\vartheta(u - u')/du' = -\delta(u - u')$ where $\delta(u - u')$ is the Dirac delta function singular at $u = u'$, demonstrates that $B_0(u)$ satisfies the differential equation

$$\ddot{B}_0 + \frac{1}{2} |W|^2 B_0 = 0. \tag{2.148}$$

Also assuming $B_\lambda(u)$ has an expansion for small $\lambda > 0$ of the Isaacson (1968a) form

$$B_\lambda(u) = B_0(u) + \lambda f_1\left(\frac{u}{\lambda}\right) B_1(u) + \lambda^2 f_2\left(\frac{u}{\lambda}\right) B_2(u) + \cdots, \tag{2.149}$$

then using (2.142) and (2.148) we find that, for small $\lambda > 0$,

$$B_\lambda(u) = B_0(u) - \frac{1}{8} \lambda^2 B_0(u) \, |W(u)|^2 \cos \frac{2u}{\lambda} + O(\lambda^3). \tag{2.150}$$

Now the line-element (2.136) becomes, for small $\lambda > 0$,

$$ds^2 = d\tilde{s}^2 + O(\lambda), \tag{2.151}$$

where

$$d\tilde{s}^2 = 2 B_0(u)^2 \, |d\zeta|^2 - 2 \, du \, dv. \tag{2.152}$$

Hence the assumption of high-frequency or short-wavelength gravitational waves has resulted in the space–time model splitting into a background space–time with line-element (2.152) and a small perturbation of first order in λ. The background space–time is not a vacuum space–time. Calculation of the Ricci tensor components $\tilde{R}_{ij}$ in the coordinates $x^i = (\zeta, \bar{\zeta}, u, v)$ using the metric given via the line-element (2.152) together with (2.148) results in

$$\tilde{R}_{ij} = |W(u)|^2 k_i\, k_j, \qquad (2.153)$$

where $k_i\, dx^i = du$ and thus k^i is a null vector field in this background space–time. The dependence on λ in (2.141) and (2.143) and the algebraic form of the background Ricci tensor (2.153) are what one expects in general for high-frequency gravitational waves following the pioneering work of Isaacson (1968a, 1968b) [see also Choquet-Bruhat (1969) and MacCallum and Taub (1973)]. For an application of these ideas to inhomogeneous plane waves see Barrabès and Hogan (2007).

Gravitational waves from isolated sources have almost spherical wavefronts asymptotically. In the simplest cases the wavefronts have histories in space–time which are expanding, shear-free null hypersurfaces (Bondi *et al.* 1962, Sachs 1962, Newman and Unti 1962, Hogan and Trautman 1987). The space–time model of the vacuum gravitational field of such waves *in the high-frequency approximation* is described by a line-element of the form (Futamase and Hogan, 1993)

$$ds^2 = 2\, r^2 p_\lambda^{-2} \left| d\zeta + \frac{\lambda\, p_\lambda^2}{r}\, \bar{W}(\bar{\zeta}, u)\, \sin\frac{u}{\lambda}\, d\bar{\zeta} \right|^2 - 2\, du\, dr - c_\lambda\, du^2. \qquad (2.154)$$

Here $p_\lambda(\zeta, \bar{\zeta}, u)$ is a real-valued function, W is an arbitrary analytic function, and the real-valued function $c_\lambda(\zeta, \bar{\zeta}, u, r)$ is given by

$$c_\lambda = K_\lambda - 2\, r\, H_\lambda - \frac{2\, m_\lambda}{r}, \qquad (2.155)$$

where

$$K_\lambda = \Delta_\lambda \log p_\lambda,\quad H_\lambda = p_\lambda^{-1} \dot{p}_\lambda,\quad m_\lambda = m_\lambda(u), \qquad (2.156)$$

with

$$\Delta_\lambda = 2\, p_\lambda^2 \frac{\partial^2}{\partial\zeta\, \partial\bar{\zeta}}, \qquad (2.157)$$

and the dot indicates partial differentiation with respect to u. Now the Ricci tensor calculated with the metric tensor given via the line-element (2.154) has the form (2.141) for small λ provided the following field equation is satisfied:

$$\dot{m}_\lambda - 3\, m_\lambda\, H_\lambda - \frac{1}{4}\Delta_\lambda K_\lambda + p_\lambda^4 |W|^2 \sin^2\frac{u}{\lambda} = 0. \qquad (2.158)$$

The corresponding Riemann curvature tensor has Newman–Penrose components $\Psi_A = O(\lambda^0)$ for $A = 0, 1, 2, 3$ and

$$\Psi_4 = \frac{1}{r}\lambda^{-1} p_\lambda^2 W \sin\frac{u}{\lambda} + O(\lambda^0) = O(\lambda^{-1}). \qquad (2.159)$$

Thus for small λ the gravitational field described by this space–time is type N in the Petrov classification with degenerate principal null direction $\partial/\partial r$. The integral curves of this vector field generate the null hypersurfaces $u = $ constant and these null geodesics have real expansion r^{-1} and complex shear

$$\sigma = \frac{\lambda p_\lambda^2}{r^2} W \sin \frac{u}{\lambda} + O(\lambda^2) = O(\lambda). \tag{2.160}$$

Thus for small λ the integral curves of $\partial/\partial r$ are expanding, shear-free null geodesics. In this sense the high-frequency waves are approximately spherical since their histories in space–time are approximately future null cones. Now writing (2.158) as

$$\frac{\partial}{\partial u}(p_\lambda^{-3} m_\lambda) = \frac{1}{4} p_\lambda^{-3} \Delta_\lambda K_\lambda - p_\lambda |W|^2 \sin^2 \frac{u}{\lambda}, \tag{2.161}$$

and multiplying it by the step function, in the same manner as we treated (2.142) above, and then taking the limit $\lambda \to 0$ using the Riemann–Lebesgue theorem, we find that p_0 and m_0 satisfy the differential equation

$$\dot{m}_0 - 3 m_0 H_0 - \frac{1}{4} \Delta_0 K_0 + \frac{1}{2} p_0^4 |W|^2 = 0. \tag{2.162}$$

Now the line-element (2.154) can be written in the form

$$ds^2 = d\tilde{s}^2 + O(\lambda), \tag{2.163}$$

with

$$d\tilde{s}^2 = 2 r^2 p_0^{-2} |d\zeta|^2 - 2\, du\, dr - c_0\, du^2, \tag{2.164}$$

with c_0 given by (2.155)–(2.157) with $\lambda = 0$. This background space–time is a Robinson–Trautman (1960, 1962) space–time with Ricci tensor components $\tilde{R}_{ij}$ in coordinates $x^i = (\zeta, \bar{\zeta}, r, u)$ given, on account of (2.162), by

$$\tilde{R}_{ij} = \frac{p_0^4 |W|^2}{r^2} k_i k_j, \tag{2.165}$$

with $k_i\, dx^i = du$. Once again the space–time model of the gravitational field of the high-frequency waves has split into a background and a small perturbation with the background Ricci tensor having a characteristic algebraic form (proportional to the square of the null propagation vector of the radiation). The background space–time is *exact* and in the present case is a non-vacuum Robinson–Trautman (1960, 1962) space–time. Robinson–Trautman space–times satisfying (2.165) are known to decay to the Schwarzschild space–time under reasonable conditions of smoothness on the 2-dimensional subspaces $u = $ constant, $r = $ constant (Lukács et al. 1984). When this happens $W = 0$ and the high-frequency radiation disappears. Such a decay of high-frequency radiation from an isolated gravitating system is not surprising.

3
Equations of motion

We describe a recently developed approach (Futamase et al. 2008, Asada et al. 2010) to obtaining equations of motion of Schwarzschild, Reissner–Nordström, or Kerr particles (with small mass and charge) moving in external fields using Einstein's vacuum field equations, or the Einstein–Maxwell vacuum field equations as appropriate, together with the assumption that near the particle the wavefronts of the radiation produced by the motion of the particle are smoothly deformed spheres. No divergent integrals arise in this approach.

3.1 Motivation

We begin with the Eddington–Finkelstein form of the Schwarzschild line-element:

$$ds^2 = p_0^{-2}(d\xi^2 + d\eta^2) - 2\, du\, dr - \left(1 - \frac{2m}{r}\right) du^2, \tag{3.1}$$

with

$$p_0 = 1 + \frac{1}{4}(\xi^2 + \eta^2), \tag{3.2}$$

and m is the constant mass of the source. The coordinates ξ, η are stereographic coordinates on the unit 2-sphere and thus have the ranges $-\infty < \xi < +\infty$, $-\infty < \eta < +\infty$. The coordinate u is a null coordinate (in the sense that the hypersurfaces $u = $ constant are null) and has the range $-\infty < u < +\infty$. The coordinate r is an affine parameter along the generators of the null hypersurfaces $u = $ constant, with ξ, η labelling these generators, and we take r to have the range $0 \leq r < +\infty$.

When $m = 0$, (3.1) becomes the line-element of Minkowskian space–time:

$$ds_0^2 = p_0^{-2}(d\xi^2 + d\eta^2) - 2\, du\, dr - du^2. \tag{3.3}$$

We will refer to this space–time as the *background space–time* and take it as a model of the external field in which the mass m is located. Since in this case it is Minkowskian space–time, no external gravitational field exists. The coordinate transformation

$$x = -r\, p_0^{-1} \xi, \tag{3.4}$$
$$y = -r\, p_0^{-1} \eta, \tag{3.5}$$

$$z = -r\, p_0^{-1}\left(1 - \frac{1}{4}(\xi^2 + \eta^2)\right), \tag{3.6}$$

$$t = u + r, \tag{3.7}$$

results in

$$ds_0^2 = dx^2 + dy^2 + dz^2 - dt^2. \tag{3.8}$$

We see from (3.4)–(3.7) that in this Minkowskian space–time $r = 0$ is the time-like geodesic $x = y = z = 0$, $t = u$ with u proper-time or arc length along it. We also see from (3.4)–(3.7) that, for $r > 0$,

$$x^2 + y^2 + z^2 = (t - u)^2 \text{ with } t - u > 0. \tag{3.9}$$

It thus follows that $u = $ constant are future null-cones with vertices on the time-like geodesic $r = 0$. Finally we observe from (3.4)–(3.7) that when $u = $ constant the coordinate r is an affine parameter along the generators of the future null-cones and that each generator is labelled by ξ, η. We will introduce the (small) mass m as a perturbation of this background space–time which is singular on $r = 0$. In the present case this perturbation is

$$\gamma_{ij}\, dx^i\, dx^j = \frac{2m}{r}\, du^2. \tag{3.10}$$

It is fortuitous in this case that the perturbation is exact in the sense that the perturbed space–time is an exact solution of Einstein's vacuum field equations. When the background space–time is non-flat the best we shall be able to achieve is knowledge of the background space–time in the neighbourhood of an arbitrary time-like world line $r = 0$ and then to obtain the perturbation of the background due to the presence of the mass (or charged mass, as the case might be) by solving approximately Einstein's vacuum field equations (or the Einstein–Maxwell vacuum field equations, if the small mass m has a small charge e). The differential equations for the time-like world line $r = 0$ in the background space–time (on which the perturbations are singular) will be called *the equations of motion* of the small mass. In the example under consideration these are the time-like geodesic equations. The same conclusion can be drawn if, instead of starting with the Schwarzschild solution (3.1), we were to have started with the Reissner–Nordström solution given by the line-element

$$ds^2 = p_0^{-2}(d\xi^2 + d\eta^2) - 2\, du\, dr - \left(1 - \frac{2m}{r} - \frac{e^2}{r^2}\right) du^2, \tag{3.11}$$

giving the space–time model of the field of a mass m of charge e. The corresponding potential 1-form is

$$A = \frac{e}{r}\, du, \tag{3.12}$$

and the corresponding Maxwell field, together with the metric given via the line-element (3.11), satisfies the vacuum Einstein–Maxwell field equations. The background space–time (got by putting $m = e = 0$) is again (3.3) and there is no external field (electromagnetic or gravitational) and so the world line $r = 0$ in the background space–time is a time-like geodesic. We thus in particular see that there is no *runaway* motion.

30 Equations of motion

To illustrate how Einstein's vacuum field equations can determine the equations of motion of a small mass m we can take a slightly broader view of the construction described above. In the background Minkowskian space–time with line-element (3.8) make, in place of (3.4)–(3.7), the coordinate transformation [see, for example, Newman and Unti (1963) or Synge (1970)]:

$$x^i = w^i(u) + r\, k^i, \qquad (3.13)$$

where $x^i = (x, y, z, t)$ for $i = 1, 2, 3, 4$. Here $r = 0$ is an arbitrary time-like world line with parametric equations $x^i = w^i(u)$. We write the components of the tangent to this line as $v^i(u) = dw^i/du$ and require this to be a unit vector so that $v_j\, v^j = \eta_{ij} v^i v^j = -1$. Thus u is proper-time or arc length along the time-like world line and $v^i(u)$ is the 4-velocity of the particle with world line $r = 0$. Its 4-acceleration is then $a^i(u) = dv^i/du$ and hence $a_j\, v^j = 0$. In addition to (3.13) we take

$$k_j\, k^j = 0 \quad \text{and} \quad k_j\, v^j = -1. \qquad (3.14)$$

Hence k^i is a null vector and it is future-pointing (since $k_j\, v^j < 0$) and normalized by the second equation here. We shall take

$$P_0\, k^i = -\xi\, \delta_1^i - \eta\, \delta_2^i - \left(1 - \frac{1}{4}(\xi^2 + \eta^2)\right)\delta_3^i + \left(1 + \frac{1}{4}(\xi^2 + \eta^2)\right)\delta_4^i. \qquad (3.15)$$

Now the second equation in (3.14) implies that

$$P_0 = \xi\, v^1(u) + \eta\, v^2(u) + \left(1 - \frac{1}{4}(\xi^2 + \eta^2)\right) v^3(u) + \left(1 + \frac{1}{4}(\xi^2 + \eta^2)\right) v^4(u). \qquad (3.16)$$

Differentiating (3.15) with respect to u yields

$$\frac{\partial k^i}{\partial u} = -h_0\, k^i, \qquad (3.17)$$

where $h_0 = P_0^{-1}\dot{P}_0$ and the dot denotes partial differentiation with respect to u. Taking the scalar product of (3.17) with respect to the Minkowskian metric tensor with components η_{ij}, and using the second of (3.14), we arrive at

$$h_0 = P_0^{-1}\dot{P}_0 = a_j\, k^j. \qquad (3.18)$$

Equation (3.17) is a transport law for k^i along the world line $r = 0$ which preserves (3.14). An alternative transport law for k^i preserving (3.14) is described in Appendix B. From (3.13) we have

$$dx^i = (v^i - r\, h_0\, k^i)\, du + k^i\, dr + r\, \frac{\partial k^i}{\partial \xi}\, d\xi + r\, \frac{\partial k^i}{\partial \eta}\, d\eta. \qquad (3.19)$$

Direct calculation from (3.15) reveals that

$$\frac{\partial k_i}{\partial \xi}\frac{\partial k^i}{\partial \xi} = \frac{\partial k_i}{\partial \eta}\frac{\partial k^i}{\partial \eta} = P_0^{-2} \quad \text{and} \quad \frac{\partial k_i}{\partial \xi}\frac{\partial k^i}{\partial \eta} = 0. \qquad (3.20)$$

Hence using (3.19) we arrive at the line-element of Minkowskian space–time:

$$ds_0^2 = \eta_{ij}\, dx^i\, dx^j = r^2 P_0^{-2}(d\xi^2 + d\eta^2) - 2\, du\, dr - (1 - 2\, h_0 r)\, du^2. \qquad (3.21)$$

If the world line $r = 0$ is a geodesic then $a^i = 0$ and we can take, without loss of generality, $v^i = \delta_4^i$ which results in $P_0 = p_0$ in (3.2) and so (3.21) reduces to (3.3) in this case. If we now introduce the mass m in the same way as it enters (3.1), just for illustrative purposes, then we are to consider a space–time with line-element

$$ds^2 = r^2 P_0^{-2}(d\xi^2 + d\eta^2) - 2\, du\, dr - \left(1 - 2 h_0 r - \frac{2m}{r}\right) du^2, \tag{3.22}$$

$$= (\vartheta^1)^2 + (\vartheta^2)^2 - 2\, \vartheta^3\, \vartheta^4, \tag{3.23}$$

where the 1-forms are given by

$$\vartheta^1 = r\, P_0^{-1} d\xi, \tag{3.24}$$

$$\vartheta^2 = r\, P_0^{-1} d\eta, \tag{3.25}$$

$$\vartheta^3 = dr + \frac{1}{2}\left(1 - 2 h_0 r - \frac{2m}{r}\right) du, \tag{3.26}$$

$$\vartheta^4 = du\, . \tag{3.27}$$

A calculation of the Ricci tensor components R_{ab} on the half-null tetrad defined via these 1-forms reveals that

$$R_{ab} = \frac{6\, m\, h_0}{r^2} \delta_a^4\, \delta_b^4. \tag{3.28}$$

Hence we see that if the vacuum field equations $R_{ab} = 0$ are to be satisfied for $m \neq 0$ we must have $h_0 = 0$ which, on account of (3.18), means that $a_j k^j = 0$ for all k^j and thus $a_j = 0$ and the world line $r = 0$ in the Minkowskian background space–time with line-element (3.21) must be a geodesic. The point of this artificial example is to demonstrate the possibility that the field equations can lead to the equations of motion in the sense that we have defined them following (3.10) above. In practice the field equations alone are not sufficient to determine the equations of motion. In addition to the field equations we will assume that the wavefronts of the radiation produced by the moving mass are smoothly deformed 2-spheres near the particle.

The calculations given in the previous paragraph were aimed at making a point regarding how the equations of motion of a small mass m might be obtained using the vacuum field equations. Up to and including (3.21) they will be extremely useful to us in the remainder of this chapter. We can take them a little further in a way that will be very helpful later. We first note that the coordinates ξ, η, r, u each depend upon the rectangular Cartesian coordinates and time $x^i = (x, y, z, t)$ in a way that is obtained by inverting the transformation (3.13). Without explicitly inverting (3.13) we can still derive the derivatives of ξ, η, r, u with respect to x^i by first differentiating (3.13) with respect to x^j to obtain

$$\delta_j^i = (v^i - r\, h_0 k^i)\, u_{,j} + k^i\, r_{,j} + r\, \frac{\partial k^i}{\partial \xi} \xi_{,j} + r\, \frac{\partial k^i}{\partial \eta} \eta_{,j}, \tag{3.29}$$

with the comma indicating partial differentiation. Multiplying this by k_i yields

$$k_j = -u_{,j}, \tag{3.30}$$

which shows that the null vector field with components k^i is tangent to the hypersurfaces $u = $ constant. Next multiplying (3.29) by v_i gives

$$v_j = -(1 - r\, h_0) u_{,j} - r_{,j}, \tag{3.31}$$

from which we have

$$r_{,j} = -v_j + (1 - r\, h_0)\, k_j. \tag{3.32}$$

Multiplying (3.29) by $\partial k_i / \partial \xi$ and using (3.20) results in

$$\xi_{,j} = \frac{1}{r} P_0^2 \frac{\partial k_j}{\partial \xi}, \tag{3.33}$$

and multiplying (3.29) by $\partial k_i / \partial \eta$ and using (3.20) results in

$$\eta_{,j} = \frac{1}{r} P_0^2 \frac{\partial k_j}{\partial \eta}. \tag{3.34}$$

Thus with $x^i = (x, y, z, t)$ we have, for future reference,

$$du = -k_j\, dx^j, \tag{3.35}$$

$$dr = -v_j\, dx^j + (1 - r\, h_0)\, k_j\, dx^j, \tag{3.36}$$

$$r\, P_0^{-1} d\xi = P_0 \frac{\partial k_j}{\partial \xi}\, dx^j, \tag{3.37}$$

$$r\, P_0^{-1} d\eta = P_0 \frac{\partial k_j}{\partial \eta}\, dx^j. \tag{3.38}$$

Now substituting (3.30), (3.32), (3.33), and (3.34) into (3.29), and raising the index j using η^{ij}, we arrive at the very useful formula:

$$\eta^{ij} = -v^i\, k^j - v^j\, k^i + k^i\, k^j + P_0^2 \left(\frac{\partial k^i}{\partial \xi} \frac{\partial k^j}{\partial \xi} + \frac{\partial k^i}{\partial \eta} \frac{\partial k^j}{\partial \eta} \right). \tag{3.39}$$

With k^i given by (3.15) we thus have

$$k_{i,j} = \frac{\partial k_i}{\partial u} u_{,j} + \frac{\partial k_i}{\partial \xi} \xi_{,j} + \frac{\partial k_i}{\partial \eta} \eta_{,j},$$

$$= h_0\, k_i\, k_j + \frac{1}{r} P_0^2 \left(\frac{\partial k_i}{\partial \xi} \frac{\partial k_j}{\partial \xi} + \frac{\partial k_i}{\partial \eta} \frac{\partial k_j}{\partial \eta} \right),$$

$$= \frac{1}{r} \{ \eta_{ij} + v_i\, k_j + v_j\, k_i - (1 - r\, h_0)\, k_i\, k_j \}. \tag{3.40}$$

We can calculate directly from (3.40) that

$$k_{i,j}\, k^j = 0, \tag{3.41}$$

$$\theta = \frac{1}{2} k^i{}_{,i} = \frac{1}{r}, \tag{3.42}$$

$$\omega = \sqrt{\frac{1}{2}k_{[i,j]}\,k^{i,j}} = 0, \tag{3.43}$$

$$|\sigma| = \sqrt{\frac{1}{2}k_{(i,j)}\,k^{i,j} - \theta^2} = 0, \tag{3.44}$$

where the square brackets denote skew-symmetrization and the round brackets denote symmetrization. The first of these confirms that the integral curves of k^i are null geodesics. The second equation shows that neighbouring null geodesics *converge* on the time-like world line $r = 0$. The third and fourth equations show that the null geodesic integral curves of k^i are *twist-free* [since they generate the null hypersurfaces $u =$ constant on account of (3.30)] and *shear-free*, respectively. Hence the null hypersurfaces $u =$ constant are future null-cones with vertices on $r = 0$. Finally we note the following second partial derivatives [see Asada *et al.* (2010) p. 85 for a derivation]:

$$\frac{\partial^2 k^i}{\partial \xi^2} = P_0^{-2}(v^i - k^i) - \frac{\partial}{\partial \xi}(\log P_0)\frac{\partial k^i}{\partial \xi} + \frac{\partial}{\partial \eta}(\log P_0)\frac{\partial k^i}{\partial \eta}, \tag{3.45}$$

$$\frac{\partial^2 k^i}{\partial \eta^2} = P_0^{-2}(v^i - k^i) + \frac{\partial}{\partial \xi}(\log P_0)\frac{\partial k^i}{\partial \xi} - \frac{\partial}{\partial \eta}(\log P_0)\frac{\partial k^i}{\partial \eta}, \tag{3.46}$$

$$\frac{\partial^2 k^i}{\partial \xi \partial \eta} = -\frac{\partial}{\partial \eta}(\log P_0)\frac{\partial k^i}{\partial \xi} - \frac{\partial}{\partial \xi}(\log P_0)\frac{\partial k^i}{\partial \eta}. \tag{3.47}$$

3.2 Example of a background space–time

The simplest example to illustrate a violation of geodesic motion is probably to consider a Reissner–Nordström particle of small mass $m = O_1$ and small charge $e = O_1$ moving in an external Einstein–Maxwell vacuum field. The external field is modelled by a general solution of the Einstein–Maxwell vacuum field equations in which an arbitrary time-like world line is identified. In the neighbourhood of this world line the space–time is Minkowskian, if we neglect $O(r^2)$ terms where r is a measure of distance from the world line. In general the line-element of the background space–time can be written in the form [derived in Asada *et al.* (2010)]

$$ds^2 = (\vartheta^1)^2 + (\vartheta^2)^2 - 2\,\vartheta^3 \vartheta^4 = g_{ab}\,\vartheta^a\,\vartheta^b, \tag{3.48}$$

with

$$\vartheta^1 = r\,p^{-1}(e^\alpha \cosh\beta\,d\xi + e^{-\alpha}\sinh\beta\,d\eta + a\,du) = \vartheta_1, \tag{3.49}$$

$$\vartheta^2 = r\,p^{-1}(e^\alpha \sinh\beta\,d\xi + e^{-\alpha}\cosh\beta\,d\eta + b\,du) = \vartheta_2, \tag{3.50}$$

$$\vartheta^3 = -dr - \frac{c}{2}\,du = -\vartheta_4, \tag{3.51}$$

$$\vartheta^4 = -du = -\vartheta_3. \tag{3.52}$$

34 Equations of motion

The constants g_{ab} are the components of the metric tensor calculated on the tetrad defined via these 1-forms. In addition the functions $p, \alpha, \beta, a, b, c$ have the form

$$p = P_0(1 + q_2\, r^2 + q_3\, r^3 + \cdots), \tag{3.53}$$

$$\alpha = \alpha_2\, r^2 + \alpha_3\, r^3 + \cdots, \tag{3.54}$$

$$\beta = \beta_2\, r^2 + \beta_3\, r^3 + \cdots, \tag{3.55}$$

$$a = a_1\, r + a_2\, r^2 + \cdots, \tag{3.56}$$

$$b = b_1\, r + b_2\, r^2 + \cdots, \tag{3.57}$$

$$c = 1 - 2\, h_0\, r + c_2\, r^2 + \cdots, \tag{3.58}$$

with P_0 and h_0 given by (3.16) and (3.18), respectively. The coefficients of the various powers of r here are functions of ξ, η, u. Thus the integral curves of the vector field $\partial/\partial r$ are null geodesics generating the null hypersurfaces $u = $ constant and have complex shear σ and expansion θ given by

$$\sigma = 2\,(\alpha_2 + i\beta_2)\, r + 3\,(\alpha_3 + i\beta_3)\, r^2 + \cdots, \tag{3.59}$$

$$\theta = \frac{1}{r} - 2\, q_2 r - 3\, q_3 r^2 + \cdots. \tag{3.60}$$

Thus the null hypersurfaces $u = $ constant are approximately future null-cones, with vertices on $r = 0$, in the vicinity of the time-like world line $r = 0$. The potential 1-form near the world line $r = 0$ can be put in the form

$$A = L\, d\xi + M\, d\eta + K\, du, \tag{3.61}$$

with

$$L = r^2\, L_2 + r^3\, L_3 + \cdots, \tag{3.62}$$

$$M = r^2\, M_2 + r^3\, M_3 + \cdots, \tag{3.63}$$

$$K = r\, K_1 + r^2\, K_2 + \cdots, \tag{3.64}$$

and with the coefficients of the powers of r functions of ξ, η, u. The tetrad components F_{ab} of the Maxwell field that we will require are

$$F_{13} = -2\, P_0\, L_2 + O(r), \tag{3.65}$$

$$F_{23} = -2\, P_0\, M_2 + O(r), \tag{3.66}$$

$$F_{34} = K_1 + O(r). \tag{3.67}$$

We thus obtain the functions L_2, M_2, and K_1 by calculating these equations on the world line $r = 0$. In view of (3.35)–(3.38) and (3.49)–(3.58) we see that when $r = 0$ we have

$$\vartheta^1 = P_0\, \frac{\partial k_j}{\partial \xi}\, dx^j = \vartheta_1, \tag{3.68}$$

$$\vartheta^2 = P_0 \frac{\partial k_j}{\partial \eta} \, dx^j = \vartheta_2, \tag{3.69}$$

$$\vartheta^3 = \left(v_j - \frac{1}{2} k_j\right) dx^j = -\vartheta_4, \tag{3.70}$$

$$\vartheta^4 = k_j \, dx^j = -\vartheta_3. \tag{3.71}$$

Thus we can conclude from (3.65)–(3.67) that

$$L_2 = \frac{1}{2} F_{ij}(u) \, k^i \, \frac{\partial k^j}{\partial \xi}, \tag{3.72}$$

$$M_2 = \frac{1}{2} F_{ij}(u) \, k^i \, \frac{\partial k^j}{\partial \eta}, \tag{3.73}$$

$$K_1 = F_{ij}(u) \, k^i \, v^j, \tag{3.74}$$

where $F_{ij}(u) = -F_{ji}(u)$ are the components of the (external) Maxwell field calculated in the coordinates $x^i = (x, y, z, t)$ on the world line $r = 0$. Maxwell's vacuum field equations require these three functions to satisfy

$$K_1 = P_0^2 \left(\frac{\partial L_2}{\partial \xi} + \frac{\partial M_2}{\partial \eta} \right), \tag{3.75}$$

$$\frac{\partial K_1}{\partial \xi} = -2 L_2 + \frac{\partial}{\partial \eta} \left\{ P_0^2 \left(\frac{\partial M_2}{\partial \xi} - \frac{\partial L_2}{\partial \eta} \right) \right\}, \tag{3.76}$$

$$\frac{\partial K_1}{\partial \eta} = -2 M_2 - \frac{\partial}{\partial \xi} \left\{ P_0^2 \left(\frac{\partial M_2}{\partial \xi} - \frac{\partial L_2}{\partial \eta} \right) \right\}, \tag{3.77}$$

and using the formulae (3.45)–(3.47) it is straightforward to verify that this is indeed the case. In addition we note that, from (3.45) and (3.46),

$$\Delta k^i = 2 \, (v^i - k^i), \tag{3.78}$$

where

$$\Delta = P_0^2 \left(\frac{\partial^2}{\partial \xi^2} + \frac{\partial^2}{\partial \eta^2} \right), \tag{3.79}$$

is the Laplacian on the unit 2-sphere. It thus follows from (3.74) that

$$\Delta K_1 + 2 K_1 = 0, \tag{3.80}$$

and so K_1 is an $l = 1$ spherical harmonic.

The tetrad components C_{abcd} of the Weyl conformal curvature tensor which we shall require in the neigbourhood of $r = 0$ are found to be

$$C_{1313} + i C_{1323} = 6 \, (\alpha_2 + i \beta_2) + O(r), \tag{3.81}$$

$$C_{3431} + i C_{3432} = \frac{3}{2} P_0^{-1} \left(a_1 + i b_1 + 4 P_0^2 \, \frac{\partial q_2}{\partial \bar{\zeta}} \right) + O(r), \tag{3.82}$$

with $\zeta = \xi + i\eta$. The functions $\alpha_2, \beta_2, a_1, b_1, q_2$ must satisfy the Einstein–Maxwell vacuum field equations

$$R_{ab} = 2\, E_{ab}, \qquad (3.83)$$

where R_{ab} are the components of the Ricci tensor calculated on the tetrad given via the 1-forms (3.49)–(3.52) and E_{ab} are the electromagnetic energy–momentum tensor components on this tetrad calculated from the tetrad components F_{ab} of the Maxwell tensor via the formula

$$E_{ab} = F_{ca}\, F^c{}_b - \frac{1}{4} g_{ab}\, F_{cd}\, F^{cd}, \qquad (3.84)$$

where the constants g_{ab}, which appear above in (3.48), are the tetrad components of the metric tensor and tetrad indices are raised and lowered using g^{ab} and g_{ab}, respectively, with $g^{ab} g_{bc} = \delta^a_c$. The field equations (3.83) yield

$$q_2 = \frac{2}{3} P_0^2 (L_2^2 + M_2^2). \qquad (3.85)$$

Substituting for L_2 and M_2 from (3.72) and (3.73) and using (3.39) results in

$$q_2 = -\frac{1}{6} F^p{}_i(u)\, F_{pj}(u)\, k^i k^j. \qquad (3.86)$$

The differential equations emerging from (3.83) to be satisfied by $\alpha_2, \beta_2, a_1, b_1$, and q_2 are

$$2(\alpha_2 + i\beta_2) = -\frac{\partial}{\partial \bar\zeta}\left(a_1 + ib_1 + 4 P_0^2 \frac{\partial q_2}{\partial \bar\zeta}\right), \qquad (3.87)$$

$$a_1 + ib_1 + 4 P_0^2 \frac{\partial q_2}{\partial \bar\zeta} = 2 P_0^4 \frac{\partial}{\partial \zeta}\left(P_0^{-2}(\alpha_2 + i\beta_2)\right). \qquad (3.88)$$

Now using (3.68)–(3.71) and (3.86) we find from (3.81) and (3.82) that

$$\alpha_2 = \frac{1}{6} P_0^2\, C_{ijkl}(u)\, k^i\, \frac{\partial k^j}{\partial \xi}\, k^k\, \frac{\partial k^l}{\partial \xi}, \qquad (3.89)$$

$$\beta_2 = \frac{1}{6} P_0^2\, C_{ijkl}(u)\, k^i\, \frac{\partial k^j}{\partial \xi}\, k^k\, \frac{\partial k^l}{\partial \eta}, \qquad (3.90)$$

and

$$a_1 = \frac{2}{3} P_0^2 \left(C_{ijkl}(u) k^i\, v^j\, k^k\, \frac{\partial k^l}{\partial \xi} + F^p{}_i(u)\, F_{pj}(u)\, k^i\, \frac{\partial k^j}{\partial \xi}\right), \qquad (3.91)$$

$$b_1 = \frac{2}{3} P_0^2 \left(C_{ijkl}(u) k^i\, v^j\, k^k\, \frac{\partial k^l}{\partial \eta} + F^p{}_i(u)\, F_{pj}(u)\, k^i\, \frac{\partial k^j}{\partial \eta}\right). \qquad (3.92)$$

Here $C_{ijkl}(u)$ are the components of the Weyl conformal curvature tensor of this background space–time calculated on $r = 0$ in the coordinates $x^i = (x, y, z, t)$. Using the formulas (3.45)–(3.47) one can verify that (3.86) along with (3.89)–(3.92) satisfy the differential equations (3.87) and (3.88).

3.3 Equations of motion of a Reissner–Nordström particle in first approximation

We now introduce the small mass m with small charge e as a perturbation of the background space–time which is singular on the world line $r=0$ in this background. The perturbed space–time will have line-element of the form (3.50)–(3.52) and the perturbed potential 1-form will have the form (3.61) but with the functions $p, \alpha, \beta, a, b, c$ given by

$$p = \hat{P}_0 \left(1 + \hat{q}_2\, r^2 + \hat{q}_3\, r^3 + \cdots \right), \tag{3.93}$$

$$\alpha = \hat{\alpha}_2\, r^2 + \hat{\alpha}_3\, r^3 + \cdots \tag{3.94}$$

$$\beta = \hat{\beta}_2\, r^2 + \hat{\beta}_3\, r^3 + \cdots, \tag{3.95}$$

$$a = \frac{\hat{a}_{-1}}{r} + \hat{a}_0 + \hat{a}_1\, r + \hat{a}_2\, r^2 + \cdots, \tag{3.96}$$

$$b = \frac{\hat{b}_{-1}}{r} + \hat{b}_0 + \hat{b}_1\, r + \hat{b}_2\, r^2 + \cdots, \tag{3.97}$$

$$c = \frac{e^2}{r^2} - \frac{2\,(m + 2\hat{f}_{-1})}{r} + \hat{c}_0 + \hat{c}_1\, r + \hat{c}_2\, r^2 + \cdots, \tag{3.98}$$

where the coefficients of the various powers of r are functions of ξ, η, u; functions here that are non-vanishing in the background differ from their background values by O_1 terms and functions here that vanish in the background are small of the following orders: $\hat{a}_{-1} = O_1$, $\hat{a}_0 = O_1$, $\hat{b}_{-1} = O_1$, $\hat{b}_0 = O_1$, and $\hat{f}_{-1} = O_2$. In addition the functions L, M, K are given by

$$L = \hat{L}_0 + r^2\, \hat{L}_2 + r^3\, \hat{L}_3 + \cdots, \tag{3.99}$$

$$M = \hat{M}_0 + r^2\, \hat{M}_2 + r^3\, \hat{M}_3 + \cdots, \tag{3.100}$$

$$K = \frac{(e + \hat{K}_{-1})}{r} + \hat{K}_0 + r\, \hat{K}_1 + r^2\, \hat{K}_2 + \cdots, \tag{3.101}$$

with

$$\hat{L}_0 = O_2, \quad \hat{M}_0 = O_2, \quad \hat{K}_0 = -e\, h_0 + O_2, \tag{3.102}$$

which ensures that near $r = 0$ the potential 1-form will resemble the Liénard–Wiechert potential 1-form up to a gauge term. Otherwise $\hat{K}_{-1} = O_2$ and the remaining coefficients of powers of r differ from their background values by O_1 terms. Finally we note that since $\hat{P}_0$ in (3.93) differs from its background value of P_0 given in (3.16) by O_1 terms we can write

$$\hat{P}_0 = P_0\,(1 + Q_1 + Q_2 + O_3), \tag{3.103}$$

where $Q_1 = O_1$ and $Q_2 = O_2$. It now remains to require that the perturbed metric tensor and potential 1-form should satisfy the perturbed Maxwell and Einstein vacuum field equations. For the purpose of the current illustration we will derive the equations of motion of the Reissner–Nordström particle of small mass and charge, moving in the external field described in the previous section, with an O_2 error. Hence we will not have to satisfy the perturbed field equations to a high degree of accuracy. For example from the perturbed Maxwell equations we shall only require $\hat{K}_1 = K_1 + O_1$ with K_1 given by (3.74). Einstein's field equations $\hat{R}_{13} - 2\,\hat{E}_{13} = 0$ and $\hat{R}_{23} - 2\,\hat{E}_{23} = 0$ have leading terms in powers of r starting with r^{-2}. When the coefficients of r^{-2} are required to be small of second order we find that

$$\hat{a}_{-1} = -4\,e\,P_0^2\,L_2 + O_2 = O_1, \tag{3.104}$$

$$\hat{b}_{-1} = -4\,e\,P_0^2\,M_2 + O_2 = O_1. \tag{3.105}$$

The leading term in $\hat{R}_{11} + \hat{R}_{22} - 2\,(\hat{E}_{11} + \hat{E}_{22}) = 0$ in powers of r is r^{-2} and requiring its coefficient to be small of second order results in

$$\hat{c}_0 = 1 + \Delta Q_1 + 2\,Q_1 + 8\,e\,F_{ij}\,k^i\,v^j + O_2. \tag{3.106}$$

Using these results in the leading r^{-2} term in $\hat{R}_{44} - 2\,\hat{E}_{44} = 0$ leads to the following differential equation for Q_1:

$$-\frac{1}{2}\Delta(\Delta Q_1 + 2\,Q_1) = 6\,m\,a_i\,p^i - 6\,e\,F_{ij}\,p^i\,v^j + O_2. \tag{3.107}$$

We have used $p^i = h^i_j\,k^j$ with $h^i_j = \delta^i_j + v^i\,v_j$ the projection tensor which projects vectors orthogonal to v^i. Thus $p^i\,v_i = 0$, $p^i\,p_i = 1$, and $p^i = k^i - v^i$. Also using (3.45) and (3.46) implies that $\Delta p^i + 2\,p^i = 0$ so that the components p^i are each $l = 1$ spherical harmonics. Integrating (3.107) and discarding the singular solution of the homogeneous equation [(3.107) with zero on the right-hand side] results in

$$\Delta Q_1 + 2\,Q_1 = 6\,m\,a_i\,p^i - 6\,e\,F_{ij}\,p^i\,v^j + A(u) + O_2, \tag{3.108}$$

where $A(u) = O_1$ is a function of integration. Now the first two terms on the right-hand side of this equation are $l = 1$ spherical harmonics and so if we require that Q_1 be singularity free as a function of ξ, η, for $-\infty < \xi < +\infty$ and $-\infty < \eta < +\infty$, we must have

$$m\,a_i\,p^i - e\,F_{ij}\,p^i\,v^j = O_2, \tag{3.109}$$

for *any* unit space-like vector p^i orthogonal to v^i. Hence we must have

$$m\,a_i = e\,F_{ij}\,v^j + O_2, \tag{3.110}$$

which are the *equations of motion in first approximation*. We have here the expected appearance of the Lorentz 4-force.

In the perturbed space–time the line-elements induced on the null hypersurfaces $u = \text{constant}$ are given, for small r, by

$$ds_0^2 = r^2\,\hat{P}_0^{-2}(d\xi^2 + d\eta^2), \tag{3.111}$$

with $\hat{P}_0$ given by (3.103). These null hypersurfaces are the histories of the possible wavefronts of the radiation produced by the motion of the charged mass. If there is no perturbation the degenerate line-elements (3.111) are the line-elements of 2-spheres.

In general we assume that they are smooth perturbations of 2-spheres in the sense that the functions Q_1, Q_2, etc. are non-singular for $-\infty < \xi < +\infty$ and $-\infty < \eta < +\infty$. Such singularities would constitute 'directional singularities' and their presence would contradict the notion that the charged mass is an isolated body. The example given above demonstrates the importance of this assumption for the derivation of the equations of motion. When the calculations given here are extended to the next order of approximation, electromagnetic radiation reaction effects appear in the equations of motion as well as 'tail terms' (which depend on the past history of the charged mass) and an additional external 4-force of second order [see Asada et al. (2010) for details].

3.4 Background space–time for a Kerr particle

We will now consider the extension of this scheme to derive equations of motion of small *rotating* masses. Equations of motion are derived which are linear in the spin of a Kerr particle moving in an external vacuum gravitational field. They exhibit spin–curvature interaction and their derivation assumes that *the spin is finite and the mass of the Kerr particle is small*. This case raises important challenges for future research. The calculations have been carried out by Shinpei Ogawa and one of us (PH). We begin by modifying the Minkowskian line-element (3.21).

To describe a spinning body we develop the geometrical construction leading to (3.21) in such a way that spin variables are introduced into the line-element of Minkowskian space–time to accompany the 4-velocity and 4-acceleration variables already present. We do this in such a way that if $a^i = 0$ then the line-element of Minkowskian space–time coincides with the Kerr line-element with three components of angular momentum (Barrabès and Hogan 2003b, p. 37) in the special case in which the mass m vanishes. Let $s^i(u)$ be the components of a vector field in coordinates $x^i = (x, y, z, t)$ defined along the world line $x^i = w^i(u)$ in Minkowskian space–time and satisfying

$$s^i v_i = 0 \text{ and } \frac{ds^i}{du} = (a_j s^j) v^i. \qquad (3.112)$$

Thus s^i is a space-like vector field which is Fermi transported along $x^i = w^i(u)$. When the mass parameter m is introduced later we will be able to interpret s^i as describing angular momentum per unit mass and so we will henceforth refer to s^i as *the spin vector*. To introduce the spin vector into the Minkowskian line-element we modify (3.13) to read

$$x^i = w^i(u) + r k^i + U^i, \qquad (3.113)$$

with

$$U^i = P_0^2 \left(\frac{\partial k^i}{\partial \xi} F_\eta - \frac{\partial k^i}{\partial \eta} F_\xi \right), \qquad (3.114)$$

$$F = -s_i k^i = -s^1 k^1 - s^2 k^2 - s^3 k^3 + s^4 k^4. \qquad (3.115)$$

40 Equations of motion

The subscripts on F denote partial derivatives and k^i and P_0 are given by (3.15) and (3.16). Clearly $U^i(\xi, \eta, u)$ satisfies $v_i U^i = 0 = k_i U^i$. Hence in particular $r = v_i(x^i - w^i(u))$. Notwithstanding appearances we emphasize that the set of points of Minkowskian space–time which lie on the world line $x^i = w^i(u)$ and the set of points of Minkowskian space–time with coordinates (3.113) with $s^i \neq 0$ are two disjoint sets. The dependence of ξ, η, r, u on x^i implied by (3.113) is different from that implied by (3.13). This will become obvious below when we calculate the partial derivatives of ξ, η, r, u with respect to x^i, which we will require later. The calculation of $\eta_{ij}\, dx^i\, dx^j$ is lengthy and makes use of the following derivatives:

$$\frac{\partial U^i}{\partial \xi} = (v^i - k^i)F_\eta + \frac{\partial k^i}{\partial \eta}F, \tag{3.116}$$

$$\frac{\partial U^i}{\partial \eta} = -(v^i - k^i)F_\xi - \frac{\partial k^i}{\partial \xi}F, \tag{3.117}$$

$$\frac{\partial U^i}{\partial u} = P_0^2\left(\frac{\partial h_0}{\partial \xi}\frac{\partial k^i}{\partial \eta} - \frac{\partial h_0}{\partial \eta}\frac{\partial k^i}{\partial \xi}\right)F$$

$$- P_0^2\left(\frac{\partial h_0}{\partial \xi}F_\eta - \frac{\partial h_0}{\partial y}F_\xi\right)k^i. \tag{3.118}$$

We also require the scalar products, with respect to the Minkowskian metric tensor η_{ij} in coordinates x^i, involving derivatives of U^i with respect ξ, η, u which are listed in Appendix C. The line-element of Minkowskian space–time can then be written in coordinates ξ, η, r, u in the form

$$ds^2 = \eta_{ij}\, dX^i\, dX^j = g_{AB}(dx^A + b^A\, d\Sigma)(dx^B + b^B\, d\Sigma) - 2dr\, d\Sigma - c\, d\Sigma^2, \tag{3.119}$$

where $x^A = (\xi, \eta)$, $b^A = g^{AB} b_B$, with g^{AB} the components of the inverse of g_{AB}, and

$$d\Sigma = -du - F_\eta\, d\xi + F_\xi\, d\eta, \tag{3.120}$$

which is not in general an exact differential,

$$g_{11} = P_0^{-2}\left(r + P_0^2 FF_\eta \frac{\partial h_0}{\partial \eta}\right)^2 + P_0^{-2}F^2\left(1 - P_0^2 F_\eta \frac{\partial h_0}{\partial \xi}\right)^2, \tag{3.121}$$

$$g_{22} = P_0^{-2}\left(r + P_0^2 FF_\xi \frac{\partial h_0}{\partial \xi}\right)^2 + P_0^{-2}F^2\left(1 + P_0^2 F_\xi \frac{\partial h_0}{\partial \eta}\right)^2, \tag{3.122}$$

$$g_{12} = g_{21} = -\frac{\partial h_0}{\partial \eta}FF_\xi\left(r + P_0^2 FF_\eta \frac{\partial h_0}{\partial \eta}\right) - \frac{\partial h_0}{\partial \xi}FF_\eta\left(r + P_0^2 FF_\xi \frac{\partial h_0}{\partial \xi}\right)$$

$$+ F^2\left(F_\xi\frac{\partial h_0}{\partial \xi} - F_\eta\frac{\partial h_0}{\partial \eta}\right), \tag{3.123}$$

$$b_1 = -r\frac{\partial}{\partial \eta}(h_0 F) - P_0^2\left(\frac{\partial h_0}{\partial \xi}F_\eta - \frac{\partial h_0}{\partial \eta}F_\xi\right)F_\eta + F_\eta$$

$$+ F^2\frac{\partial h_0}{\partial \xi} - F^2 P_0^2 F_\eta\left(\left(\frac{\partial h_0}{\partial \xi}\right)^2 + \left(\frac{\partial h_0}{\partial \eta}\right)^2\right), \quad (3.124)$$

$$b_2 = r\frac{\partial}{\partial \xi}(h_0 F) + P_0^2\left(\frac{\partial h_0}{\partial \xi}F_\eta - \frac{\partial h_0}{\partial \eta}F_\xi\right)F_\xi - F_\xi$$

$$+ F^2\frac{\partial h_0}{\partial \eta} + F^2 P_0^2 F_\xi\left(\left(\frac{\partial h_0}{\partial \xi}\right)^2 + \left(\frac{\partial h_0}{\partial \eta}\right)^2\right), \quad (3.125)$$

and

$$c = 1 - 2h_0 r - 2P_0^2\left(\frac{\partial h_0}{\partial \xi}F_\eta - \frac{\partial h_0}{\partial \eta}F_\xi\right)$$

$$- F^2 P_0^2\left(\left(\frac{\partial h_0}{\partial \xi}\right)^2 + \left(\frac{\partial h_0}{\partial \eta}\right)^2\right) + b_A b^A. \quad (3.126)$$

We note that if the 4-acceleration a^i vanishes then $h_0 = 0$ and we can choose $v^i(u) = \delta_4^i$ for all u and the line-element (3.119) reduces to

$$ds^2 = (r^2 + F^2)p_0^{-2}(d\xi^2 + d\eta^2) - 2d\Sigma\left\{dr + \frac{1}{2}(du - F_\eta\, d\xi + F_\xi\, d\eta)\right\}, \quad (3.127)$$

with p_0 defined as in (3.2). The Kerr solution with mass m and angular momentum $\mathbf{J} = (ms^1, ms^2, ms^3)$ is given by (Barrabès and Hogan, 2003b)

$$ds^2 = (r^2 + F^2)p_0^{-2}(d\xi^2 + d\eta^2) - 2d\Sigma\{dr - F_\eta\, d\xi + F_\xi\, d\eta + S\, d\Sigma\}, \quad (3.128)$$

with

$$S = \frac{1}{2} - \frac{m\, r}{r^2 + F^2}. \quad (3.129)$$

Hence in (3.127) we have arrived at a form of the line-element of Minkowskian space–time which is obtained from the Kerr solution, with three components of angular momentum, specialized to the case $m = 0$.

Clearly the line-element (3.119) is very complicated. To simplify matters *we shall henceforth neglect spin–spin terms and in addition assume that the 4-acceleration a^i is proportional to the spin s^i*. If there is no spin then the question we are considering becomes that of deriving the equations of motion of a Schwarzschild particle moving in an external vacuum gravitational field. This leads to geodesic motion in first approximation (see Section 3.3 when the charge vanishes). Hence we expect violation of geodesic motion due to spin to be proportional to the spin in first approximation. Thus h_0 is proportional to s^i and (3.119) reduces to

42 Equations of motion

$$ds^2 = r^2 P_0^{-2}\left\{\left(d\xi + \frac{P_0^2}{r^2}F_\eta\, d\Sigma\right)^2 + \left(d\eta - \frac{P_0^2}{r^2}F_\xi\, d\Sigma\right)^2\right\}$$
$$-2\, dr\, d\Sigma - (1 - 2h_0 r)\, d\Sigma^2, \tag{3.130}$$

with $d\Sigma$ defined by (3.120). The spin–spin terms here can be neglected. We have included them simply to make the expression for the line-element easier to work with below when introducing basis 1-forms [see (3.149)–(3.152)].

The coordinate transformation (3.113) expresses the rectangular Cartesian and time coordinates x^i in terms of the coordinates ξ, η, r, u. Conversely we can in principle express ξ, η, r, u each as functions of x^i and equivalently therefore consider ξ, η, r, u as scalar functions on Minkowskian space–time. To calculate the gradients of these functions we begin by differentiating (3.113) with respect to x^j to obtain, neglecting spin–spin terms,

$$\delta^i_j = (v^i - r\, h_0 k^i)u_{,j} + k^i r_{,j} + \left(r\frac{\partial k^i}{\partial \xi} + \frac{\partial U^i}{\partial \xi}\right)\xi_{,j} + \left(r\frac{\partial k^i}{\partial \eta} + \frac{\partial U^i}{\partial \eta}\right)\eta_{,j}. \tag{3.131}$$

Multiplying this successively by k_i, v_i, $\partial k_i/\partial \xi$, and $\partial k_i/\partial \eta$, and utilizing the scalar products in Appendix C, we arrive at

$$k_{,j} = -u_{,j} - F_\eta\, \xi_{,j} + F_\xi\, \eta_{,j}, \tag{3.132}$$

$$v_{,j} = -(1 - rh_0)u_{,j} - r_{,j}, \tag{3.133}$$

$$\frac{\partial k_j}{\partial \xi} = rP_0^{-2}\xi_{,j} + P_0^{-2} F\, \eta_{,j}, \tag{3.134}$$

$$\frac{\partial k_j}{\partial \eta} = -P_0^{-2} F\, \xi_{,j} + rP_0^{-2}\eta_{,j}, \tag{3.135}$$

respectively. Solving these, neglecting spin–spin terms, we find that

$$\xi_{,j} = \frac{P_0^2}{r}\frac{\partial k_j}{\partial \xi} - \frac{P_0^2 F}{r^2}\frac{\partial k_j}{\partial \eta}, \tag{3.136}$$

$$\eta_{,j} = \frac{P_0^2}{r}\frac{\partial k_j}{\partial \eta} + \frac{P_0^2 F}{r^2}\frac{\partial k_j}{\partial \xi}, \tag{3.137}$$

$$r_{,j} = -v_j + (1 - r\, h_0)k_j + \frac{F_\eta P_0^2}{r}\frac{\partial k_j}{\partial \xi} - \frac{F_\xi P_0^2}{r}\frac{\partial k_j}{\partial \eta}, \tag{3.138}$$

$$u_{,j} = -k_j - \frac{F_\eta P_0^2}{r}\frac{\partial k_j}{\partial \xi} + \frac{F_\xi P_0^2}{r}\frac{\partial k_j}{\partial \eta}. \tag{3.139}$$

When these are substituted into (3.131), neglecting spin–spin terms, we obtain the Minkowskian metric tensor components in coordinates x^i in the expected form

$$\eta^{ij} = P_0^2\left(\frac{\partial k^i}{\partial \xi}\frac{\partial k^j}{\partial \xi} + \frac{\partial k^i}{\partial \eta}\frac{\partial k^j}{\partial \eta}\right) - k^i v^j - k^j v^i + k^i k^j. \tag{3.140}$$

Using (3.136)–(3.139) we can derive the formulae:

$$k_{i,j} = \frac{1}{r}\{\eta_{ij} + k_i v_j + v_i k_j - (1 - r h_0) k_i k_j\}$$
$$+ \frac{F P_0^2}{r^2}\left(\frac{\partial k_i}{\partial \xi}\frac{\partial k_j}{\partial \eta} - \frac{\partial k_i}{\partial \eta}\frac{\partial k_j}{\partial \xi}\right), \tag{3.141}$$

$$U_{i,j} = -\frac{1}{r}(v_i - k_i) U_j + \frac{F P_0^2}{r}\left(\frac{\partial k_i}{\partial \xi}\frac{\partial k_j}{\partial \eta} - \frac{\partial k_i}{\partial \eta}\frac{\partial k_j}{\partial \xi}\right), \tag{3.142}$$

and in addition (3.138) can be written

$$r_{,j} = -v_j + (1 - r h_0) k_j + \frac{1}{r} U_j. \tag{3.143}$$

From (3.138) we have $r_{,j} k^j = -1$ and, as in the spin–free case above, we can take r as a parameter along the integral curves of k^i. As a consequence of (3.141) we have the equations

$$\frac{\partial k^i}{\partial r} = k^i{}_{,j} k^j = 0, \quad \frac{\partial k^i}{\partial u} = k^i{}_{,j} v^j = -h_0 k^i, \tag{3.144}$$

satisfied in this case also. The first of these shows that the integral curves of k^i, in the present case with $s^i \neq 0$, are null geodesics with r an affine parameter along them. The second equation in (3.144) is consistent with k^i given by (3.15) and (3.16). We easily calculate from (3.141) that, neglecting spin–spin terms,

$$\theta = \frac{1}{2} k^i{}_{,i} = \frac{1}{r}, \tag{3.145}$$

$$\omega = \sqrt{\frac{1}{2} k_{[i,j]} k^{i,j}} = \frac{F}{r^2}, \tag{3.146}$$

$$|\sigma| = \sqrt{\frac{1}{2} k_{(i,j)} k^{i,j} - \theta^2} = 0, \tag{3.147}$$

with the square brackets around indices denoting skew-symmetrization and the round brackets denoting symmetrization. These equations mean that, neglecting spin–spin terms, the integral curves of the null vector field k^i which satisfies (3.141) are geodesics having expansion θ, twist ω, and vanishing complex shear σ (with $|\sigma|$ denoting the modulus of σ). Using (3.136)–(3.139) we find that $k^i \partial/\partial X^i = \partial/\partial r$ confirming that r is a parameter along the integral curves of k^i.

Writing (3.130) in terms of basis 1-forms we have

$$ds^2 = (\vartheta^1)^2 + (\vartheta^2)^2 - 2\vartheta^3 \vartheta^4, \tag{3.148}$$

with

$$\vartheta^1 = r P_0^{-1}\left(d\xi + \frac{P_0^2 F_\eta}{r^2} du\right) = \vartheta_1, \tag{3.149}$$

$$\vartheta^2 = r P_0^{-1}\left(d\eta - \frac{P_0^2 F_\xi}{r^2} du\right) = \vartheta_2, \tag{3.150}$$

44 *Equations of motion*

$$\vartheta^3 = dr + \frac{1}{2}(1 - 2h_0 r)\, d\Sigma = -\vartheta_4, \tag{3.151}$$

$$\vartheta^4 = d\Sigma = -\vartheta_3. \tag{3.152}$$

By (3.136)–(3.139) these can be written

$$\vartheta_1 = \left(P_0 \frac{\partial k_j}{\partial \xi} + \frac{P_0 F}{r} \frac{\partial k_j}{\partial \eta} - \frac{P_0 F_\eta}{r} k_j \right) dx^j, \tag{3.153}$$

$$\vartheta_2 = \left(P_0 \frac{\partial k_j}{\partial \eta} - \frac{P_0 F}{r} \frac{\partial k_j}{\partial \xi} + \frac{P_0 F_\xi}{r} k_j \right) dx^j, \tag{3.154}$$

$$\vartheta_3 = k_j\, dx^j, \tag{3.155}$$

$$\vartheta_4 = \left(v_j - \frac{1}{2} k_j - \frac{P_0^2 F_\eta}{r} \frac{\partial k_j}{\partial \xi} + \frac{P_0^2 F_\xi}{r} \frac{\partial k_j}{\partial \eta} \right) dx^j. \tag{3.156}$$

These will be of particular interest to us in the next section when we discuss the full background space–time.

Defining the vector field

$$K^i = k^i + \frac{1}{r} U^i, \tag{3.157}$$

using (3.139) and (3.141)–(3.143) we see that, neglecting spin–spin terms, K^i satisfies

$$K^i K_i = 0, \qquad K^i v_i = -1, \tag{3.158}$$

$$K_j = -u_{,j}, \tag{3.159}$$

and

$$K_{i,j} = \frac{1}{r}\{\eta_{ij} + K_i v_j + v_i K_j - (1 - r h_0) K_i K_j\}. \tag{3.160}$$

Thus K^i satisfies the same equations, and therefore possesses the same geometrical properties, as k^i does when $s^i = 0$ [comparing (3.160) with (3.40)]. In particular K^i is approximately (neglecting spin–spin terms) null, geodesic, twist-free, shear-free, and has expansion r^{-1}. By (3.159) the integral curves of K^i generate the approximately null hypersurfaces $u = $ constant. By (3.113) we have

$$x^i = w^i(u) + r K^i, \tag{3.161}$$

and so for large values of r we see that K^i approximates k^i and that the null geodesic integral curves of K^i appear to converge on the world line $x^i = w^i(u)$. For this reason in the sequel we will regard $x^i = w^i(u)$ as representing the history of the relativistically moving Kerr particle and the equations of motion, which we seek using Einstein's vacuum field equations, are differential equations for this world line. Finally we note that

$$P_0^2 \left(\frac{\partial k^i}{\partial \xi} \frac{\partial k^j}{\partial \eta} - \frac{\partial k^i}{\partial \eta} \frac{\partial k^j}{\partial \xi} \right) = \epsilon^{ijkl} k_k v_l, \tag{3.162}$$

where ϵ^{ijkl} is the 4-dimensional Levi-Città permutation symbol with the convention that $\epsilon^{1234} = 1$ and (3.114) can thus be written

$$U^i = \epsilon^{ijkl} s_j k_k v_l. \tag{3.163}$$

Defining the *spin tensor*

$$s_{ij} = \epsilon_{ijkl} v^k s^l = -s_{ji}, \tag{3.164}$$

we have $s_{ij} v^j = 0$ and, using the properties of the permutation symbol,

$$s^i = \frac{1}{2}\epsilon^{ijkl} v_j s_{kl}, \tag{3.165}$$

demonstrating that the spin vector s^i is the Hodge dual of the spin tensor contracted once with the 4-velocity v^i. Substitution of (3.165) into (3.163) yields

$$U^i = s^{ij} k_j, \tag{3.166}$$

and thus (3.157) can be written

$$K^i = \left(\eta^{ij} + \frac{1}{r}s^{ij}\right) k_j, \tag{3.167}$$

indicating that, for large r, K^i differs from k^i by an infinitesimal Lorentz transformation generated by the spin tensor s^{ij}.

3.5 Equations of motion of a Kerr particle in first approximation

In the manner adopted in Section 3.2 we require a general form for the line-element of the background space–time, which plays the role of the space–time model of the external vacuum gravitational field. In the light of Section 3.2 and of (3.120) we write the line-element of the background space–time in the form

$$ds^2 = r^2 p^{-2}\Big\{(e^\alpha \cosh\beta\, d\xi + e^{-\alpha} \sinh\beta\, d\eta + a\, d\Sigma)^2$$
$$+(e^\alpha \sinh\beta\, d\xi + e^{-\alpha} \cosh\beta\, d\eta + b\, d\Sigma)^2\Big\} - 2\, dr\, d\Sigma - c\, d\Sigma^2,$$
$$= (\theta^1)^2 + (\theta^2)^2 - 2\theta^3 \theta^4, \tag{3.168}$$

with

$$\theta^1 = r\, p^{-1}(e^\alpha \cosh\beta\, d\xi + e^{-\alpha} \sinh\beta\, d\eta + a\, d\Sigma), \tag{3.169}$$
$$\theta^2 = r\, p^{-1}(e^\alpha \sinh\beta\, d\xi + e^{-\alpha} \cosh\beta\, d\eta + b\, d\Sigma), \tag{3.170}$$
$$\theta^3 = -dr + \frac{1}{2}c\, d\Sigma, \tag{3.171}$$
$$\theta^4 = d\Sigma, \tag{3.172}$$

and

$$p = P_0(1 + q_1 r + q_2 r^2 + \cdots), \tag{3.173}$$

$$\alpha = \alpha_1 r + \alpha_2 r^2 + \cdots, \tag{3.174}$$

$$\beta = \beta_1 r + \beta_2 r^2 + \cdots, \tag{3.175}$$

$$a = \frac{1}{r^2} P_0^2 F_\eta + \frac{a_{-1}}{r} + a_0 + a_1 r + \cdots, \tag{3.176}$$

$$b = -\frac{1}{r^2} P_0^2 F_\xi + \frac{b_{-1}}{r} + b_0 + b_1 r + \cdots, \tag{3.177}$$

$$c = c_0 + c_1 r + \cdots. \tag{3.178}$$

The coefficients of the various powers of r here are functions of ξ, η, and u. We can obtain $\alpha_1, \alpha_2, \beta_1, \beta_2$ by consideration of the Riemann curvature tensor of the background. We shall require some of the tetrad components of the background curvature tensor R_{abcd}. To obtain these we return to the Minkowskian space–time considered in the previous section. Using the tetrad defined via the 1-forms (3.153)–(3.156) the tetrad components which are of particular interest to us are R_{1313} and R_{1323}. When written in coordinates x^i these are found to be

$$R_{1313} = \frac{2}{r} P_0^2 F R_{ijkl}(u) \frac{\partial k^i}{\partial \xi} k^j \frac{\partial k^k}{\partial \eta} k^l + P_0^2 R_{ijkl}(u) \frac{\partial k^i}{\partial \xi} k^j \frac{\partial k^k}{\partial \xi} k^l$$

$$+ O(F) + O(r), \tag{3.179}$$

$$R_{1323} = -\frac{2}{r} P_0^2 F R_{ijkl}(u) \frac{\partial k^i}{\partial \xi} k^j \frac{\partial k^k}{\partial \xi} k^l + P_0^2 R_{ijkl}(u) \frac{\partial k^i}{\partial \xi} k^j \frac{\partial k^k}{\partial \eta} k^l$$

$$+ O(F) + O(r), \tag{3.180}$$

where $R_{ijkl}(u)$ are the components of the Riemann curvature tensor of the background space–time, in coordinates x^i, calculated on the world line $x^i = w^i(u)$. In arriving at (3.179) and (3.180) we have Taylor expanded $R_{ijkl}(x^p)$ about the world line $x^i = w^i(u)$ using (3.113) and neglected spin–spin terms. It is important to emphasize that the world line $x^i = w^i(u)$ does *not* correspond to $r = 0$ on account of (3.113). Direct calculation using (3.168)–(3.179) yields in this case

$$R_{1313} = \frac{2(\alpha_1 + 2F\beta_2)}{r} + 6\alpha_2 + 6F\beta_3 + O(r), \tag{3.181}$$

$$R_{1323} = \frac{2(\beta_1 - 2F\alpha_2)}{r} + 6\beta_2 - 6F\alpha_3 + O(r), \tag{3.182}$$

from which we conclude, in the light of (3.179) and (3.180), that

$$\alpha_1 = 4F\beta_2, \tag{3.183}$$

$$\beta_1 = -4F\alpha_2, \tag{3.184}$$

$$\alpha_2 = \frac{1}{6} P_0^2 R_{ijkl}(u) \frac{\partial k^i}{\partial \xi} k^j \frac{\partial k^k}{\partial \xi} k^l + O(F), \tag{3.185}$$

$$\beta_2 = \frac{1}{6} P_0^2 R_{ijkl}(u) \frac{\partial k^i}{\partial \xi} k^j \frac{\partial k^k}{\partial \eta} k^l + O(F). \tag{3.186}$$

Equations of motion of a Kerr particle in first approximation 47

The spin terms, denoted by $O(F)$ here, will not be required in the sequel since terms arising there which involve α_2 and β_2 will be multiplied by spin terms and we consistently neglect spin–spin terms. If R_{ab} denotes the components of the background Ricci tensor calculated on the half-null tetrad defined by the 1-forms (3.169)–(3.179) then these components in general contain a finite number of terms involving inverse powers of r followed by a term independent of r and an infinite number of terms involving positive powers of r. The background space–time is a vacuum space–time and so $R_{ab} = 0$. Equating to zero the coefficients of r^{-1} and r^0 in R_{33} gives

$$q_1 = 0, \qquad q_2 = 0. \tag{3.187}$$

The vanishing of the leading r^0 terms in R_{13} and R_{23} yield

$$a_1 = P_0^4 \left(\frac{\partial}{\partial \xi}(P_0^{-2}\alpha_2) + \frac{\partial}{\partial \eta}(P_0^{-2}\beta_2) \right) + O(F),$$

$$= \frac{2}{3} P_0^2 R_{ijkl}(u) k^i v^j k^k \frac{\partial k^l}{\partial \xi} + O(F), \tag{3.188}$$

$$b_1 = -P_0^4 \left(\frac{\partial}{\partial \eta}(P_0^{-2}\alpha_2) - \frac{\partial}{\partial \xi}(P_0^{-2}\beta_2) \right) + O(F),$$

$$= \frac{2}{3} P_0^2 R_{ijkl}(u) k^i v^j k^k \frac{\partial k^l}{\partial \eta} + O(F). \tag{3.189}$$

The vanishing of the r^0 terms in $R_{11} - R_{22}$ and R_{12} result in

$$\alpha_2 = \frac{1}{4}\left(-\frac{\partial a_1}{\partial \xi} + \frac{\partial b_1}{\partial \eta}\right) + O(F), \tag{3.190}$$

$$\beta_2 = \frac{1}{4}\left(-\frac{\partial a_1}{\partial \eta} - \frac{\partial b_1}{\partial \xi}\right) + O(F), \tag{3.191}$$

and these are satisfied by the functions given above in (3.185), (3.186), (3.188), and (3.189). Next the vanishing of the coefficient of r^{-1} in $R_{11} - R_{22} + 2i R_{12}$ provides the complex differential equation

$$\left(\frac{\partial}{\partial \xi} + i\frac{\partial}{\partial \eta}\right)(a_0 + ib_0) = 14\,iF(\alpha_2 + i\beta_2),$$

$$= -7i\left(\frac{\partial}{\partial \xi} + i\frac{\partial}{\partial \eta}\right)\left[(F_\xi - iF_\eta) P_0^2(\alpha_2 + i\beta_2)\right]. \tag{3.192}$$

The two real equations here are solved by the particular integrals:

$$a_0 = -7P_0^2(\alpha_2 F_\eta - \beta_2 F_\xi), \tag{3.193}$$

$$b_0 = -7P_0^2(\alpha_2 F_\xi + \beta_2 F_\eta). \tag{3.194}$$

48 Equations of motion

The vanishing of the coefficients of r^{-2} in R_{13} and R_{23} results in

$$a_{-1} = 0, \tag{3.195}$$
$$b_{-1} = 0. \tag{3.196}$$

Finally the vanishing of the coefficients of r^{-2} and r^{-1} in $R_{11} + R_{22}$ yields, respectively,

$$c_0 = P_0^2 \left(\frac{\partial^2}{\partial \xi^2} + \frac{\partial^2}{\partial \eta^2} \right) \log P_0 = \Delta \log P_0 = 1, \tag{3.197}$$

$$c_1 = -2 h_0 - FP_0^2 \left(\frac{\partial}{\partial \eta}(P_0^{-2} a_1) - \frac{\partial}{\partial \xi}(P_0^{-2} b_1) \right) + 5(a_1 F_\eta - b_1 F_\xi), \tag{3.198}$$

while the remaining Ricci tensor components involve functions which will not be required.

Following the example of Section 3.3 we now introduce the small mass m of the rotating source with 4-velocity v^i, 4-acceleration a^i, and spin s^i by perturbing the functions (3.173)–(3.179) to read

$$\hat{p} = \hat{P}_0(1 + \hat{q}_1 r + \hat{q}_2 r^2 + \cdots), \tag{3.199}$$

$$\hat{\alpha} = \hat{\alpha}_1 r + \hat{\alpha}_2 r^2 + \cdots, \tag{3.200}$$

$$\hat{\beta} = \hat{\beta}_1 r + \hat{\beta}_2 r^2 + \cdots, \tag{3.201}$$

$$\hat{a} = \frac{1}{r^2} P_0^2 F_\eta + \frac{\hat{a}_{-1}}{r} + \hat{a}_0 + \hat{a}_1 r + \cdots, \tag{3.202}$$

$$\hat{b} = -\frac{1}{r^2} P_0^2 F_\xi + \frac{\hat{b}_{-1}}{r} + \hat{b}_0 + \hat{b}_1 r + \cdots, \tag{3.203}$$

$$\hat{c} = -\frac{2 m}{r} + \hat{c}_0 + \hat{c}_1 r + \cdots, \tag{3.204}$$

where perturbed quantities, indicated with hats, differ from their background values by terms of first order which we denote by $O_1 = O(m)$. It will be convenient to write $\hat{P}_0$ in the form (3.103). The coefficients of the various powers of r here are functions of ξ, η, u. We will solve Einstein's vacuum field equations for them to the accuracy required to determine the equations of motion for the world line $x^i = w^i(u)$ described above in first approximation. In so doing we will have to ensure that the resulting components of the perturbed metric tensor are free of directional singularities. Although we have already explained how these singularities manifest themselves in the present context in Section 3.3 we emphasize again that these are singularities corresponding to particular values of the coordinates ξ and η and typically such singularities correspond to $\xi = \pm\infty$ and/or $\eta = \pm\infty$. The perturbed half-null tetrad $\hat{\vartheta}^1, \hat{\vartheta}^2, \hat{\vartheta}^3, \hat{\vartheta}^4$ is given by (3.169)–(3.172) with $p, \alpha, \beta, a, b, c$ replaced by $\hat{p}, \hat{\alpha}, \hat{\beta}, \hat{a}, \hat{b}, \hat{c}$ above. The vanishing of the perturbed Ricci tensor components $\hat{R}_{ab}$ on the perturbed half-null tetrad constitute Einstein's vacuum field equations. As in the case of the background space–time these components typically involve a finite number of terms involving inverse powers of r followed by an infinite series in r. We need to derive equations which ensure that the coefficients of these powers of r are small in terms of m. We systematically work through these coefficients making them sufficiently small in order to derive the equations of motion for the world line $x^i = w^i(u)$ in the background space–time in first approximation (with an O_2 error). To begin with we find that

$$\hat{R}_{33} = -\frac{4}{r}(\hat{q}_1 + O_2) + (6\,\hat{q}_2 + O_2) + O(r), \tag{3.205}$$

and so we take

$$\hat{q}_1 = O_2, \qquad \hat{q}_2 = O_2. \tag{3.206}$$

The coefficients of r^{-2} in $\hat{R}_{13}$ and $\hat{R}_{23}$ are small of second order in terms of m provided

$$\hat{a}_{-1} - 2\,\hat{b}_0 F + O_2, \tag{3.207}$$
$$\hat{b}_{-1} = -2\,\hat{a}_0 F + O_2. \tag{3.208}$$

We note that these equations are consistent with (3.193)–(3.196) because the zeroth order values of a_0 and b_0 given in (3.193) and (3.194) are spin terms and, since we are consistently neglecting spin–spin terms, they give no contribution to the right-hand sides of (3.207) and (3.208). Thus there are no zeroth-order terms on the right-hand sides of (3.207) and (3.208) and this is consistent with (3.195) and (3.196). The coefficient of r^{-1} in $\hat{R}_{13}$, denoted $_{(-1)}\hat{R}_{13}$, is given by

$$_{(-1)}\hat{R}_{13} = 4\,b_1 P_0^{-1} F - 4\,P_0(\alpha_2 F_\eta - \beta_2 F_\xi) - P_0^3\left(\frac{\partial}{\partial \xi}(P_0^{-2}\alpha_1) + \frac{\partial}{\partial \eta}(P_0^{-2}\beta_1)\right) + O_1. \tag{3.209}$$

With α_1, β_1 given by (3.183) and (3.184) and with a_1 given by (3.188) this reduces to

$$_{(-1)}\hat{R}_{13} = O_1, \tag{3.210}$$

neglecting, as always, spin–spin terms. This accuracy will be sufficient for our purposes. Similarly

$$_{(-1)}\hat{R}_{23} = -4\,a_1 P_0^{-1} F - 4\,P_0(-\alpha_2 F_\xi + \beta_2 F_\eta)$$
$$- P_0^3\left(-\frac{\partial}{\partial \eta}(P_0^{-2}\alpha_1) + \frac{\partial}{\partial \xi}(P_0^{-2}\beta_1)\right) + O_1,$$
$$= O_1. \tag{3.211}$$

The coefficients of r^0 in $\hat{R}_{13}$ and $\hat{R}_{23}$ are the background values plus O_1 terms. The background values vanish on account of (3.188) and (3.189) and thus these coefficients are both O_1 quantities. It will be convenient from now on to denote the coefficient of r^n in $\hat{R}_{ab}$ by $_{(n)}\hat{R}_{ab}$. With the vanishing of the background values of $_{(0)}\hat{R}_{11} - _{(0)}\hat{R}_{22}$ and $_{(0)}\hat{R}_{12}$ leading to (3.190) and (3.191) these components of the perturbed Ricci tensor are small O_1 quantities. In order to determine the O_1 terms in $\hat{a}_{-1}$ and $\hat{b}_{-1}$ we see from (3.207) and (3.208) that we require the O_1 terms in $\hat{a}_0$ and $\hat{b}_0$. Writing $\hat{a}_0 = a_0 + \mathcal{A} + O_2$ and $\hat{b}_0 = b_0 + \mathcal{B} + O_2$ with $\mathcal{A} = O_1$ and $\mathcal{B} = O_1$ we find, using the background values (3.193) and (3.194), that

50 Equations of motion

$$_{(-1)}\hat{R}_{11} - {}_{(-1)}\hat{R}_{22} = 2\left(\frac{\partial A}{\partial \xi} - \frac{\partial B}{\partial \eta}\right) - 16\, m\, \alpha_2 + O(m\,F) + O_2, \qquad (3.212)$$

$$_{(-1)}\hat{R}_{12} = \frac{\partial A}{\partial \eta} + \frac{\partial B}{\partial \xi} - 8\, m\, \beta_2 + O(m\,F) + O_2 . \qquad (3.213)$$

These are solved with the particular integrals

$$A = -2\, m\, a_1 + O(m\,F) + O_2 \quad\text{and}\quad B = -2\, m\, b_1 + O(m\,F) + O_2, \qquad (3.214)$$

and thus in the light of (3.193) and (3.194) we have

$$\hat{a}_0 = -7 P_0^2 (\alpha_2 F_\eta - \beta_2 F_\xi) - 2\, m\, a_1 + O(m\,F) + O_2, \qquad (3.215)$$

$$\hat{b}_0 = -7 P_0^2 (\alpha_2 F_\xi + \beta_2 F_\eta) - 2\, m\, b_1 + O(m\,F) + O_2, \qquad (3.216)$$

and in addition (3.207) and (3.208) give

$$\hat{a}_{-1} = -4\, m\, F\, b_1 + O_2, \qquad (3.217)$$

$$\hat{b}_{-1} = 4\, m\, F\, a_1 + O_2, \qquad (3.218)$$

after neglecting spin–spin terms. Now

$$\begin{aligned}
_{(-2)}\hat{R}_{11} - {}_{(-2)}\hat{R}_{22} &= \frac{\partial \hat{a}_{-1}}{\partial \xi} - \frac{\partial \hat{b}_{-1}}{\partial \eta} - 2\, \hat{b}_0 F_\xi - 2\, \hat{a}_0 F_\eta - 4\, m\alpha_1 + O_2 , \\
&= -4\, m\, F \left(\frac{\partial a_1}{\partial \eta} + \frac{\partial b_1}{\partial \xi} + 4 \beta_2\right) + O_2 , \\
&= O_2,
\end{aligned} \qquad (3.219)$$

by (3.191) neglecting spin–spin terms and similarly

$$\begin{aligned}
_{(-2)}\hat{R}_{12} &= \frac{1}{2}\frac{\partial \hat{a}_{-1}}{\partial \eta} + \frac{1}{2}\frac{\partial \hat{b}_{-1}}{\partial \xi} + \hat{a}_0 F_\xi - \hat{b}_0 F_\eta - 2\, m\, \beta_1 + O_2, \\
&= 2\, m\, F \left(-\frac{\partial b_1}{\partial \eta} + \frac{\partial a_1}{\partial \xi} + 4 \alpha_2\right) + O_2, \\
&= O_2,
\end{aligned} \qquad (3.220)$$

on account of (3.190) neglecting spin–spin terms. We now calculate $\hat{R}_{11} + \hat{R}_{22}$ and to begin with find that

$$\begin{aligned}
_{(-2)}\hat{R}_{11} + {}_{(-2)}\hat{R}_{22} = {}& 2\hat{c}_0 - 2\hat{\Delta} \log \hat{P}_0 + 3 P_0^2 \left(\frac{\partial}{\partial \xi}(P_0^{-2}\hat{a}_{-1}) + \frac{\partial}{\partial \eta}(P_0^{-2}\hat{b}_{-1})\right) \\
& -4\, F P_0^2 \left(\frac{\partial}{\partial \eta}(P_0^{-2}\hat{a}_0) - \frac{\partial}{\partial \xi}(P_0^{-2}\hat{b}_0)\right) \\
& -2\,(\hat{a}_0 F_\eta - \hat{b}_0 F_\xi) + O_2,
\end{aligned} \qquad (3.221)$$

where, with $\hat{P}_0$ given by (3.103), we have

$$\hat{\Delta} \log \hat{P}_0 = \hat{P}_0^2 \left(\frac{\partial^2}{\partial \xi^2} + \frac{\partial^2}{\partial \eta^2}\right) \log \hat{P}_0 = 1 + \Delta Q_1 + 2\, Q_1 + O_2. \qquad (3.222)$$

Thus $_{(-2)}\hat{R}_{11} + {}_{(-2)}\hat{R}_{22} = O_2$ yields, following substitution of $\hat{a}_{-1}, \hat{a}_0, \hat{b}_{-1}$, and $\hat{b}_0$ from above,

$$\hat{c}_0 = 1 + \Delta Q_1 + 2\,Q_1 - 10\,m\,FP_0^2\left(\frac{\partial}{\partial \eta}(P_0^{-2}a_1) - \frac{\partial}{\partial \xi}(P_0^{-2}b_1)\right)$$
$$- 8\,m\,(a_1 F_\eta - b_1 F_\xi) + O_2\,. \tag{3.223}$$

Next $_{(-1)}\hat{R}_{11} + {}_{(-2)}\hat{R}_{22} = O_1$ results in $\hat{c}_1$ given by (3.198) with an O_1 error, which will be of sufficient accuracy for our purposes. On account of (3.207) and (3.208) we have

$$_{(-2)}\hat{R}_{34} = -\frac{1}{2}P_0^2\left(\frac{\partial}{\partial \xi}(P_0^{-2}\hat{a}_{-1}) + \frac{\partial}{\partial \eta}(P_0^{-2}\hat{b}_{-1})\right) - (\hat{a}_0 F_\eta - \hat{b}_0 F_\xi)$$
$$- FP_0^2\left(\frac{\partial}{\partial \eta}(P_0^{-2}\hat{a}_0) - \frac{\partial}{\partial \xi}(P_0^{-2}\hat{b}_0)\right) + O_2$$
$$= O_2\,. \tag{3.224}$$

Since $\hat{c}_1 = c_1 + O_1$ with c_1 given by (3.198) we find that

$$_{(-1)}\hat{R}_{34} = O_1. \tag{3.225}$$

Turning now to $\hat{R}_{14}$ and $\hat{R}_{24}$ we have

$$_{(-3)}\hat{R}_{14} = P_0 F_\eta\,(\Delta Q_1 + 2\,Q_1) + O_2 = O_1, \tag{3.226}$$

and

$$_{(-3)}\hat{R}_{24} = -P_0 F_\xi\,(\Delta Q_1 + 2\,Q_1) + O_2 = O_1. \tag{3.227}$$

In addition we have enough functions determined explicitly to find that $_{(-2)}\hat{R}_{14} = O_1 = {}_{(-2)}\hat{R}_{24}$. Finally we consider $\hat{R}_{44}$. The leading term in the series of increasing powers of r is the r^{-2} term. Its coefficient is $_{(-2)}\hat{R}_{44}$ and $_{(-2)}\hat{R}_{44} = O_2$ is given explicitly by

$$6\,m\,h_0 = \Delta\left\{-\frac{1}{2}(\Delta Q_1 + 2\,Q_1) + 5\,m\,FP_0^2\left(\frac{\partial}{\partial \eta}(P_0^{-2}a_1) - \frac{\partial}{\partial \xi}(P_0^{-2}b_1)\right)\right.$$
$$\left. + 4\,m\,(a_1 F_\eta - b_1 F_\xi)\right\} + 6\,m\,P_0^2 F\left(\frac{\partial}{\partial \eta}(P_0^{-2}a_1) - \frac{\partial}{\partial \xi}(P_0^{-2}b_1)\right)$$
$$+ 9\,m\,(a_1 F_\eta - b_1 F_\xi) - 3\,mP_0^2\left(\frac{\partial}{\partial \xi}(P_0^{-2}a_0) + \frac{\partial}{\partial \eta}(P_0^{-2}b_0)\right)$$
$$+ 2\,m\,P_0^2\left(F_\xi\frac{\partial}{\partial \eta} - F_\eta\frac{\partial}{\partial \xi}\right)\left[P_0^2\left(\frac{\partial}{\partial \xi}(P_0^{-2}a_1) + \frac{\partial}{\partial \eta}(P_0^{-2}b_1)\right)\right]$$
$$+ O_2\,. \tag{3.228}$$

We note that F given by (3.115) is an $l = 1$ spherical harmonic (thus $\Delta F + 2F = 0$), $h_0 = P_0^{-1}\partial P_0/\partial u$ is also an $l = 1$ spherical harmonic, a_1 and b_1 are given by (3.188) and

52 Equations of motion

(3.189), and a_0 and b_0 are given by (3.193) and (3.194). In preparation for substitution into (3.228) we obtain

$$FP_0^2 \left(\frac{\partial}{\partial \eta}(P_0^{-2}a_1) - \frac{\partial}{\partial \xi}(P_0^{-2}b_1) \right) = 2 I_1, \tag{3.229}$$

$$P_0^2 \left(F_\xi \frac{\partial}{\partial \eta} - F_\eta \frac{\partial}{\partial \xi} \right) \left[P_0^2 \left(\frac{\partial}{\partial \xi}(P_0^{-2}a_1) + \frac{\partial}{\partial \eta}(P_0^{-2}b_1) \right) \right] = -4 I_2, \tag{3.230}$$

$$(a_1 F_\eta - b_1 F_\xi) = \frac{2}{3} I_3, \tag{3.231}$$

$$P_0^2 \left(\frac{\partial}{\partial \xi}(P_0^{-2}a_0) + \frac{\partial}{\partial \eta}(P_0^{-2}b_0) \right) = -\frac{14}{3} I_3, \tag{3.232}$$

where

$$I_1 = P_0^2 R_{ijkl}(u) s_m k^i v^j \frac{\partial k^k}{\partial \eta} \frac{\partial k^l}{\partial \xi} k^m, \tag{3.233}$$

$$I_2 = P_0^2 R_{ijkl}(u) s_m v^i k^j v^k \left(\frac{\partial k^l}{\partial \xi} \frac{\partial k^m}{\partial \eta} - \frac{\partial k^l}{\partial \eta} \frac{\partial k^m}{\partial \xi} \right), \tag{3.234}$$

$$I_3 = P_0^2 R_{ijkl}(u) s_m k^i v^j k^k \left(\frac{\partial k^l}{\partial \xi} \frac{\partial k^m}{\partial \eta} - \frac{\partial k^l}{\partial \eta} \frac{\partial k^m}{\partial \xi} \right). \tag{3.235}$$

Using (3.162) and the definition of the spin tensor (3.164) we can write

$$I_1 = \hat{I}_1 - \frac{1}{5} R_{ijkl}(u) k^i v^j s^{kl} = \Delta \left(-\frac{1}{12} \hat{I}_1 + \frac{1}{10} R_{ijkl} k^i v^j s^{kl} \right), \tag{3.236}$$

$$I_2 = \Delta \left(-\frac{1}{6} I_2 \right), \tag{3.237}$$

$$I_3 = \hat{I}_1 - I_2 + \frac{3}{10} R_{ijkl} k^i v^j s^{kl} = \Delta \left(-\frac{1}{12} \hat{I}_1 + \frac{1}{6} I_2 - \frac{3}{20} R_{ijkl} k^i v^j s^{kl} \right). \tag{3.238}$$

Here (3.236) *defines* $\hat{I}_1$ which is an $l = 3$ spherical harmonic while the Riemann tensor term in (3.238) is an $l = 1$ spherical harmonic. We see from (3.237) that I_2 is an $l = 2$ spherical harmonic while (3.238) demonstrates that I_3 is a linear combination of $l = 1, l = 2$, and $l = 3$ spherical harmonics. Substituting (3.229)–(3.232) and (3.236)–(3.238) into (3.228) makes it possible to rewrite (3.228) in the form

$$\Delta \left[-\frac{1}{2}(\Delta + 2)(Q_1 + 2m \hat{I}_1 + m I_2) + 3m h_0 - 3m R_{ijkl} k^i v^j s^{kl} \right] = O_2. \tag{3.239}$$

Integrating this equation without introducing directional singularities results in

$$\frac{1}{2}(\Delta + 2)(Q_1 + 2m \hat{I}_1 + m I_2) = 3m h_0 - 3m R_{ijkl} k^i v^j s^{kl}$$
$$+ W(u) + O_2,$$
$$= 3m a_i k^i - 3m R_{ijkl} k^i v^j s^{kl}$$
$$+ W(u) + O_2, \tag{3.240}$$

where $W(u) = O_1$ is an $l=0$ spherical harmonic. The first two terms on the right-hand side here are each $l=1$ spherical harmonics and unless they combine to produce zero (or O_2) the solution Q_1 will have directional singularities. Hence we must have

$$m \, a_i k^i = m \, R_{ijkl} k^i v^j s^{kl} + O_2, \tag{3.241}$$

for all k^i and thus we arrive at *the equations of motion in first approximation*:

$$m \, a^i = m \, R^i{}_{jkl} v^j s^{kl} + O_2. \tag{3.242}$$

In the light of (3.242) the equations of motion for the spin given in (3.112) take the form, neglecting spin–spin terms,

$$\frac{ds^i}{du} = O_1. \tag{3.243}$$

Finally we note that an $l=0$ or $l=1$ spherical harmonic in Q_1 is a trivial perturbation (Futamase et al. 2008) and so, without loss of generality, we can put $W(u) = 0$ and with the equations of motion (3.242) holding arrive at the solution

$$Q_1 = -2 \, m \, \hat{I}_1 - m \, I_2, \tag{3.244}$$

which is a linear combination of an $l=3$ and an $l=2$ spherical harmonic.

3.6 Spinning test particles

The subject of equations of motion of small spinning masses moving in external gravitational fields has a long history in general relativity starting most notably with the work of Mathisson (1937) and followed by Papapetrou (1951), Corinaldesi and Papapetrou (1951), and Dixon (1970a, 1970b, 1973). The significance of Mathisson's pioneering contribution has been assessed recently by Sauer and Trautman (2008) and by Dixon (2008). Equations of motion of the form (3.242) have been obtained but with a factor of one-half multiplying the 4-force on the right-hand side of (3.242). Such equations of motion apply to *spinning test particles* whose gravitational fields are neglected. A derivation of the equations of motion of a spinning test particle moving in an external gravitational field is given, for example, by Papapetrou (1951). A specific comparison with our approach is as follows:

(a) Papapetrou considers a small spinning mass whose gravitational field is neglected (a test particle) whereas we consider the gravitational field of the small spinning mass as a perturbation of a background space–time (not a test particle).

(b) For Papapetrou the history of the small spinning mass is a time-like world tube in a background space–time in which an energy–momentum–stress tensor T^{ab} describes physically the particle. With respect to T^{ab} multipole moments are defined and the spinning mass is technically a pole–dipole particle. We describe the spinning mass as a Kerr particle by making use of the Kerr solution.

(c) For Papapetrou and for us the background space–time is a model of the gravitational field in which the small spinning mass is moving. However Papapetrou does not use the field equations of the background whereas we do.

(d) Papapetrou obtains the equations of motion from $T^{ab}{}_{;b} = 0$, written in tensor density form, where the covariant derivative, indicated by a semicolon, is with respect to the Levi-Civita connection associated with the background metric tensor. We obtain the equations of motion using the vacuum field equations for the space–time perturbed by the presence of the small spinning mass.

Given these comparisons it would be a surprise indeed if the equations of motion we derived coincided exactly with those of Papapetrou. The algebraic form of the right-hand side of (3.242) is what one would expect but the numerical factor differing from Papapetrou's factor of one-half is a consequence of our use of the vacuum field equations, which were not used by Papapetrou.

Using the current formalism it is relatively easy to devise a strategy to obtain Papapetrou's equations of motion for a spinning *test particle*. In general relativity a test particle has a non-zero rest-mass and its gravitational field can be neglected compared to the gravitational field in which it is freely moving. By the geodesic hypothesis, a non-spinning test particle in a vacuum gravitational field has a time-like geodesic world line in the space–time model of the gravitational field in which it is moving. We ask the question: given the geodesic hypothesis for non-spinning test particles what are the equations of motion of a spinning test particle moving in a vacuum gravitational field? For simplicity we will continue to neglect spin–spin terms, but this restriction could be systematically relaxed. We first consider the non-spinning test particle whose world line, according to the geodesic hypothesis, is a time-like geodesic. From the calculations in Section 3.4 with $s^i = 0$ we find in coordinates $x^i = (\xi, \eta, r, u)$ that the only coordinate component of the metric tensor involving the 4-acceleration a^i of the world line $x^i = w^i(u)$ is

$$g_{44} = -1 + 2\, r\, h_0 + O(r^2). \tag{3.245}$$

The coefficient of $2\, r$ on the right-hand side is an $l = 1$ spherical harmonic which vanishes if, and only if, $a^i = 0$. In this case the world line $x^i = w^i(u)$ is the history of a test particle moving in the vacuum gravitational field modeled by the space–time described in Section 3.4. Equation (3.245) is a statement about the embedding of the 3-surface $u = \text{constant}$ in the 4-dimensional space–time since it represents the scalar product, with respect to the metric tensor of the space–time, of the vector field $\partial/\partial u$ with itself and $\partial/\partial u$ is part of the description of the extrinsic geometry of the 3-surface. We note in passing that the (intrinsic) Gaussian curvature of the 2-surface $r = \text{constant}, u = \text{constant}$ does not involve a^i explicitly. It is given by $\mathcal{K}/r^2$ with

$$\mathcal{K} = 1 + 2\, r^2\, R_{ijkl}(u)\, k^i\, v^j\, k^k\, v^l + O(r^2). \tag{3.246}$$

The coefficient of $2\, r^2$ here is an $l = 2$ spherical harmonic. The absence of an $l = 1$ spherical harmonic in $\mathcal{K}$ means that the 2-surface has no 'conical singularities' (Robinson and Robinson, 1972) and thus is a smoothly deformed 2-sphere.

Turning now to the test particle with spin whose world line $x^i = w^i(u)$ is described in Section 3.4, the Gaussian curvature of the 2-surface $u = \text{constant}, r = \text{constant}$ is found in this case to be $\mathcal{K}/r^2$ with

$$K = 1 + r\left(-2\,I_1 + 4\,I_2 - \frac{4}{3} I_3\right) + O(r^2), \qquad (3.247)$$

with I_1, I_2, I_3 given in (3.233)–(3.238). Substitution of (3.236)–(3.238) into (3.247) results in

$$K = 1 + r\left(\frac{16}{3} I_2 - \frac{10}{3} \hat{I}_1\right) + O(r^2). \qquad (3.248)$$

In this calculation the $l = 1$ spherical harmonic has dropped out and, as in the non-spinning case, the 2-surface $u = $ constant, $r = $ constant is again a smoothly deformed sphere. In addition (3.245) is now replaced by

$$\begin{aligned} g_{44} &= -1 + 2\,r\,(h_0 + I_1 - I_3) + O(r^2) \,, \\ &= -1 + 2\,r\left(h_0 - \frac{1}{2} R_{ijkl}(u)\,k^i\,v^j\,s^{kl} + I_2\right) + O(r^2), \end{aligned} \qquad (3.249)$$

using (3.236)–(3.238). The first two terms in the coefficient of $2\,r$ here are $l = 1$ spherical harmonics while the third term in the coefficient of $2\,r$ is an $l = 2$ spherical harmonic. In the non-spinning case the equations of motion of a test particle are obtained by equating to zero the $l = 1$ coefficient of $2\,r$ in (3.245). If the equations of motion of a spinning test particle are obtained in the same way then equating to zero the $l = 1$ terms in the coefficient of $2\,r$ in (3.249) results in

$$a_i\,k^i = h_0 = \frac{1}{2} R_{ijkl}(u)\,k^i\,v^j\,s^{kl}, \qquad (3.250)$$

for $k^i = v^i + p^i$ with p^i *any* unit space-like vector satisfying $v_i\,p^i = 0$, and $p_i\,p^i = 1$. It thus follows that the equations of motion of a spinning test particle, neglecting spin–spin terms, are

$$a^i = \frac{1}{2} R^i{}_{jkl}(u)\,v^j\,s^{kl}, \qquad (3.251)$$

in agreement with Papapetrou (1951). The calculations described here have been extended to include spin–spin terms by Bolgar (2012).

The approach to equations of motion described in this chapter offers a basis for a critical assessment of the work of, for example, Gralla and Wald (2008) and of Pound (2010) [see also the review by Poisson et al. (2011)] on the same topic. This is an important challenge for future research.

4
Inhomogeneous aspects of cosmology

The construction of inhomogeneous cosmological models in general relativity has been comprehensively reviewed by Krasiński (2006). We are concerned in this chapter with incorporating gravitational waves into isotropic cosmologies giving rise to inhomogeneities which are exact or approximate. We begin by considering exact inhomogeneity in the form of the Ozsváth–Robinson–Rózga plane-fronted gravitational waves in the presence of a cosmological constant Λ. This is followed by describing covariant and gauge-invariant perturbation theory and using it to demonstrate how gravitational waves can be introduced into isotropic cosmological models as perturbations. These gravitational waves can then be used to construct a model of cosmic background gravitational radiation.

4.1 Plane-fronted gravitational waves with a cosmological constant

The exact vacuum space–time modelling the gravitational field of plane gravitational waves, given by the line-element (2.21) with H given by the specific harmonic function (2.15), can be generalized to (2.21) with $H(x, y, u)$ a solution of

$$\frac{\partial^2 H}{\partial x^2} + \frac{\partial^2 H}{\partial y^2} = 0. \tag{4.1}$$

This generalization describes the gravitational field of plane-fronted waves with parallel rays (so-called pp waves) and is a solution of Einstein's vacuum field equations. When $H = 0$ the space–time is Minkowskian with line-element

$$ds_0^2 = dx^2 + dy^2 - 2\, du\, dv, \tag{4.2}$$

and (4.1) is the wave equation on this background space–time satisfied by any function $H(x, y, u)$. In the space–times with line-elements (2.21) or (4.2) $u =$ constant are a family of null hypersurfaces generated by the null, geodesic integral curves of the vector field $\partial/\partial v$. This vector field is covariantly constant and thus the integral curves have vanishing expansion and shear. Since they generate $u =$ constant they are obviously twist-free. Introducing a cosmological constant into solutions of Einstein's vacuum field equations, such as the Schwarzschild and Kerr solutions, is quite a simple matter. Strangely introducing a cosmological constant into what is arguably the simpler pp-wave solution is not so straightforward, as the masterly construction by Ozsváth, Robinson and Rózga (1985) demonstrates. Nevertheless it has striking similarities to the vacuum pp-wave case. The line-element can be written in the form

$$ds^2 = ds_0^2 + \Phi\, du^2, \qquad (4.3)$$

where

$$ds_0^2 = 2\,p^{-2} d\zeta\, d\bar\zeta - 2\,p^{-2} q^2\, du\, dr - p^{-2} q^2 (2\, q^{-1} \dot q\, r - \kappa(u)\, r^2)\, du^2, \qquad (4.4)$$

with

$$p = 1 + \frac{\Lambda}{6}\zeta\bar\zeta, \qquad (4.5)$$

$$q = \beta(u)\,\bar\zeta + \bar\beta(u)\,\zeta + \alpha(u)\left(1 - \frac{\Lambda}{6}\zeta\bar\zeta\right), \qquad (4.6)$$

$$\kappa(u) = 2\,\beta\bar\beta + \frac{\Lambda}{3}\,\alpha, \qquad (4.7)$$

where $\alpha(u)$ is an arbitrary real-valued function of u and $\beta(u)$ is an arbitrary complex-valued function of u (with complex conjugate denoted by a bar) and the dot indicates partial differentiation with respect to u. The line-element (4.4) is that of de Sitter space–time in a coordinate system $(\zeta, \bar\zeta, r, u)$ based on a family of null hyperplanes $u = \text{constant}$ generated by the null, geodesic integral curves of the vector field $\partial/\partial r$ which have vanishing expansion, twist, and shear (but $\partial/\partial r$ is not covariantly constant). In each null hyperplane the integral curves are labelled by $\zeta, \bar\zeta$. To complete the parallel with the vacuum pp waves the real-valued function $\Phi(\zeta, \bar\zeta, u)$ in (4.3) is a solution of the wave equation on the space–time with line-element (4.4) [or (4.3)].

We will assume that $\Lambda \neq 0$ but otherwise not distinguish between the cases $\Lambda > 0$ and $\Lambda < 0$ and thus we shall refer to all cases simply as de Sitter space–time. It is well known [see, for example Synge (1965), p. 261] that de Sitter space–time can be visualized as a quadric V_4 in a 5-dimensional flat manifold V_5. The line-element of V_5 is taken to be

$$-dl^2 = \frac{3}{\Lambda}(dX^0)^2 + (dX^1)^2 + (dX^2)^2 + (dX^3)^2 - (dX^4)^2. \qquad (4.8)$$

The equation of the quadric V_4 in this 5-dimensional manifold is taken to be

$$\frac{3}{\Lambda}(X^0)^2 + (X^1)^2 + (X^2)^2 + (X^3)^2 - (X^4)^2 = \frac{3}{\Lambda}. \qquad (4.9)$$

Following Synge it is useful to write these two equations in an obvious vector notation as

$$-dl^2 = d\mathbf{X} \cdot d\mathbf{X}, \qquad (4.10)$$

and

$$\mathbf{X} \cdot \mathbf{X} = \frac{3}{\Lambda}, \qquad (4.11)$$

respectively. We can parametrize the points on (4.11) with the four real parameters x^i with $i = 1, 2, 3, 4$ as follows:

$$X^0 = 1 - 2\lambda, \quad X^i = \lambda\, x^i, \qquad (4.12)$$

with

$$\lambda = \left(1 + \frac{\Lambda}{12}\eta_{ij}x^i x^j\right)^{-1}, \qquad (4.13)$$

with η_{ij} given via (1.1). Substituting (4.12) and (4.13) into (4.8) specializes dl^2 to

$$ds_0^2 = \frac{\eta_{ij}\, dx^i\, dx^j}{\left(1 + \frac{\Lambda}{12}\eta_{ij}x^i x^j\right)^2} = g_{ij}\, dx^i\, dx^j. \qquad (4.14)$$

Straightforward calculation of the Riemann curvature tensor components R_{ijkm} for the metric given by this line-element yields

$$R_{ijkm} = \frac{\Lambda}{3}\left(g_{im}\, g_{jk} - g_{jm}\, g_{ik}\right). \qquad (4.15)$$

This calculation confirms that (4.10) subject to the restriction (4.11) is the line-element of the de Sitter universe of constant curvature $\Lambda/3$. To obtain the form (4.4) for the line-element of de Sitter space–time we parametrize the points of V_4 using $(\zeta, \bar{\zeta}, r, u)$ by writing the position vector of a point on $V_4 \subset V_5$ in the form (Barrabès and Hogan, 2007)

$$\mathbf{X} = \mathbf{Y}(\zeta, \bar{\zeta}, u) + p^{-1} q\, r\, \mathbf{a}(u), \qquad (4.16)$$

with p, q given by (4.5) and (4.6),

$$Y^0 = p^{-1}\left(1 - \frac{\Lambda}{6}\zeta\bar{\zeta}\right), \qquad (4.17)$$

$$Y^1 + iY^2 = p^{-1}\sqrt{2}\,\zeta, \qquad (4.18)$$

$$Y^3 = p^{-1}\left\{l(u)\,\bar{\zeta} + \bar{l}(u)\,\zeta + m(u)\left(1 - \frac{\Lambda}{6}\zeta\bar{\zeta}\right)\right\} = Y^4, \qquad (4.19)$$

and

$$a^0(u) = -\frac{\Lambda}{3}\, m(u), \qquad (4.20)$$

$$a^1(u) + ia^2(u) = \sqrt{2}\, l(u), \qquad (4.21)$$

$$a^3(u) - a^4(u) = -1, \qquad (4.22)$$

$$a^3(u) + a^4(u) = \frac{\Lambda}{3}\, m^2 + 2\, l\, \bar{l}, \qquad (4.23)$$

with

$$\dot{l}(u) = \beta(u)\ \text{ and }\ \dot{m}(u) = \alpha(u). \qquad (4.24)$$

It thus follows that

$$\mathbf{Y}\cdot\mathbf{Y} = \frac{3}{\Lambda},\ \mathbf{a}\cdot\mathbf{Y} = 0,\ \mathbf{a}\cdot\mathbf{a} = 0\ \text{ and }\ \dot{\mathbf{a}}\cdot\dot{\mathbf{a}} = \kappa(u). \qquad (4.25)$$

Hence in particular $\mathbf{Y}$ is a point on V_4 corresponding to $r = 0$. Also $\mathbf{X}$ in (4.16) is a point on V_4. Substitution of (4.16) into (4.10) to obtain the induced line-element ds_0^2, given by (4.4), on V_4 requires the following scalar products:

$$\mathbf{a} \cdot \frac{\partial \mathbf{Y}}{\partial u} = -p^{-1}q, \quad \frac{\partial \mathbf{Y}}{\partial u} \cdot \frac{\partial \mathbf{Y}}{\partial u} = 0, \quad \dot{\mathbf{a}} \cdot \frac{\partial \mathbf{Y}}{\partial u} = 0, \tag{4.26}$$

and

$$\frac{\partial \mathbf{Y}}{\partial \zeta} \cdot \frac{\partial \mathbf{Y}}{\partial \bar{\zeta}} = p^{-2}, \quad \frac{\partial \mathbf{Y}}{\partial \zeta} \cdot \frac{\partial \mathbf{Y}}{\partial \zeta} = 0 = \frac{\partial \mathbf{Y}}{\partial \bar{\zeta}} \cdot \frac{\partial \mathbf{Y}}{\partial \bar{\zeta}}, \quad \dot{\mathbf{a}} \cdot \frac{\partial \mathbf{Y}}{\partial \zeta} = \frac{\partial}{\partial \zeta}(p^{-1}q). \tag{4.27}$$

Tran (1988) has shown that the intersection of the quadric V_4 with the null hyperplane passing through the origin of V_5 given by

$$\mathbf{b} \cdot \mathbf{X} = 0 \quad \text{with} \quad \mathbf{b} \cdot \mathbf{b} = 0, \tag{4.28}$$

is a null hyperplane in V_4. Moreover he has provided beautiful proofs that properties of the functions $\alpha(u), \beta(u)$ along with the sign of Λ determine whether or not the null hyperplanes $u = \text{constant}$ in V_4 intersect. His results can be summarized briefly as follows: if $\Lambda > 0$ then $\kappa > 0$ implies intersections and if $\Lambda < 0$ then (a) $\kappa < 0$ implies intersections, (b) $\kappa = 0$ implies intersections except if $\text{Im}\beta = 0$ or $\text{Re}\beta = 0$ or $\text{Re}\beta = C\,\text{Im}\beta$ for some constant C, and (c) $\kappa > 0$ implies intersections.

To complete the parallel with pp waves in the vacuum case we require the function $\Phi(\zeta, \bar{\zeta}, u)$ in (4.3) to satisfy the wave equation in the space–time with line-element (4.3) or (4.4). Thus Φ must satisfy

$$p^{-1}q\,\Phi_{\zeta\bar{\zeta}} + (p^{-1}q)_\zeta\,\Phi_{\bar{\zeta}} + (p^{-1}q)_{\bar{\zeta}}\,\Phi_\zeta = 0, \tag{4.29}$$

with the subscripts denoting partial derivatives. This can be simplified by putting

$$\Phi = p\,q^{-1}H(\zeta, \bar{\zeta}, u), \tag{4.30}$$

and using the fact that

$$(p^{-1}q)_{\zeta\bar{\zeta}} = -\frac{\Lambda}{3} p^{-3}q. \tag{4.31}$$

Thus (4.3) reads

$$ds^2 = ds_0^2 + p\,q^{-1}H\,du^2, \tag{4.32}$$

and $H(\zeta, \bar{\zeta}, u)$ satisfies

$$H_{\zeta\bar{\zeta}} + \frac{\Lambda}{3} p^{-2}H = 0. \tag{4.33}$$

The Ozsváth, Robinson, and Rózga (1985) solution of Einstein's vacuum field equations with a cosmological constant is given by (4.32) together with (4.4) and (4.33). The Weyl conformal curvature tensor of this space–time is type N in the Petrov classification with $\partial/\partial r$ as degenerate principal null direction confirming that the space–time is a model of the gravitational field due to pure gravitational radiation.

4.2 Perturbations of isotropic cosmologies

All of the exact equations required before introducing approximations can be found in the comprehensive lecture notes of Ellis (1971) with the exception that the Bianchi identities given there apply only to a perfect fluid matter distribution whereas we shall require them for a general matter distribution. The covariant approach to cosmology originated in a systematic way with the work of Schücking, Ehlers and Sachs [see Ellis (1971) and Ehlers (1993) for example] and Hawking (1966) gave the first description of cosmological perturbations in this context. We make use of the gauge-invariant and covariant cosmological perturbation theory of Ellis and Bruni (1989). This is a particularly elegant way of handling cosmological perturbations which describe matter and/or gravitational waves.

Since we are interested in space–times which are models of the gravitational field of the cosmos we have, in a general local coordinate system $\{x^a\}$, a metric tensor with components $g_{ab} = g_{ba}$ and a preferred congruence of world lines tangent to a unit time-like vector field with components u^a (with $g_{ab}\, u^a\, u^b = u_b\, u^b = -1$). The components of the Riemann curvature tensor will be denoted R_{abcd}. The Ricci tensor components are $R_{bd} = g^{ac}\, R_{abcd}$ with g^{ac} defined as usual by $g^{ac}\, g_{cb} = \delta^a_b$ and the Ricci scalar is $R = g^{ab}\, R_{ab} = R^a{}_a$. The Weyl conformal curvature tensor has components C_{abcd} given in (2.82) above. With respect to the 4-velocity field u^a the Weyl tensor is decomposed into an 'electric part' E_{ab} and a 'magnetic' part H_{ab} given by

$$E_{ab} = C_{apbq}\, u^p\, u^q \quad \text{and} \quad H_{ab} = {}^*C_{apbq}\, u^p\, u^q. \tag{4.34}$$

Here ${}^*C_{apbq} = \tfrac{1}{2}\eta_{ap}{}^{rs}\, C_{rsbq}$ are the components of the left dual of the Weyl tensor. The right dual has components $C^*_{apbq} = \tfrac{1}{2}\eta_{bq}{}^{rs}\, C_{aprs}$. In both cases $\eta_{abcd} = \sqrt{-g}\,\epsilon_{abcd}$ where $g = \det(g_{ab})$ and ϵ_{abcd} are the components of the 4-dimensional Levi-Cività permutation symbol. It is interesting to make use of the algebraic symmetries of the Weyl tensor to establish that the left and right duals are equal. Clearly the left and right duals of the Riemann curvature tensor exist and they are *not* equal, unless the space–time is a vacuum space–time. A knowledge of E_{ab} and H_{ab} is equivalent to a knowledge of C_{abcd}. The expression for the Weyl tensor components in terms of E_{ab} and H_{ab}, got by effectively inverting (4.34), is given by Ellis (1971). The energy–momentum–stress tensor describing the matter content of the universe has components $T^{ab} = T^{ba}$ which can be decomposed with respect to the vector field u^a in the form

$$T^{ab} = \mu\, u^a\, u^b + p\, h^{ab} + q^a\, u^b + q^b\, u^a + \pi^{ab}, \tag{4.35}$$

with

$$h_{ab} = g_{ab} + u_a\, u_b, \tag{4.36}$$

the projection tensor. In (4.35)

$$\mu = T_{ab}\, u^a\, u^b, \tag{4.37}$$

is interpreted physically as the matter energy density measured by the observer with 4-velocity u^a and

$$p = \frac{1}{3}\, T_{ab}\, h^{ab}, \tag{4.38}$$

is the isotropic pressure. Also
$$q^a = -T_{bc}\, h^{ba}\, u^c, \tag{4.39}$$
is the energy flow (or heat flow) measured by the observer with 4-velocity u^a and satisfying $q^a\, u_a = 0$, and
$$\pi^{ab} = T^{cd}\left(h_c^a\, h_d^b - \frac{1}{3} h^{ab}\, T^{cd}\, h_{cd}\right) = \pi^{ba}, \tag{4.40}$$
is the anisotropic stress (due, for example, to viscosity) and satisfying $\pi^{ab}\, u_b = 0 = \pi^a_a$. We indicate covariant differentiation with respect to the Levi-Cività connection associated with the metric tensor g_{ab} by a semicolon, partial differentiation by a comma, and covariant differentiation in the direction of u^a by a dot. Hence the 4-acceleration of the time-like congruence is
$$\dot{u}^a = u^a{}_{;b}\, u^b, \tag{4.41}$$
and $u_{a;b}$ can be decomposed as follows:
$$u_{a;b} = \omega_{ab} + \sigma_{ab} + \frac{1}{3}\theta\, h_{ab} - \dot{u}_a\, u_b. \tag{4.42}$$

Here
$$\omega_{ab} = u_{[a;b]} + \dot{u}_{[a}\, u_{b]} = -\omega_{ba}, \tag{4.43}$$
is the vorticity tensor of the congruence tangent to u^a. The square brackets denote skew-symmetrization. Also
$$\sigma_{ab} = u_{(a;b)} + \dot{u}_{(a}\, u_{b)} - \frac{1}{3}\theta\, h_{ab} = \sigma_{ba}, \tag{4.44}$$
is the shear tensor, with round brackets denoting symmetrization and
$$\theta = u^a{}_{;a}, \tag{4.45}$$
is the expansion (if $\theta > 0$) or contraction (if $\theta < 0$) scalar of the congruence. It thus follows from (4.44) that $g^{ab}\, \sigma_{ab} = 0$.

The equations which will play a central role in the sequel are obtained by projecting in the direction of u^a and orthogonal to u^a, using the projection tensor (4.36), the *Ricci identities*
$$u_{a;dc} - u_{a;cd} = R_{abcd}\, u^b, \tag{4.46}$$
the *equations of motion* and the *energy conservation equation* contained in
$$T^{ab}{}_{;b} = 0, \tag{4.47}$$
with T^{ab} given by (4.35), and the *Bianchi identities*
$$R_{abcd;f} + R_{abfc;d} + R_{abdf;c} = 0, \tag{4.48}$$
written in a more convenient form in terms of the Weyl conformal curvature tensor as
$$C^{abcd}{}_{;d} = R^{c[a;b]} - \frac{1}{6} g^{c[a}\, R^{;b]}. \tag{4.49}$$

Einstein's field equations, absorbing the coupling constant into the energy–momentum–stress tensor, read

$$R_{ab} - \frac{1}{2} g_{ab} R = T_{ab}. \qquad (4.50)$$

The projections mentioned above of the Ricci identities yield *Raychaudhuri's equation*,

$$\dot{\theta} + \frac{1}{3} \theta^2 - \dot{u}^a{}_{;a} + 2(\sigma^2 - \omega^2) + \frac{1}{2}(\mu + 3p) = 0, \qquad (4.51)$$

where $\sigma^2 = \frac{1}{2} \sigma_{ab} \sigma^{ab}$ and $\omega^2 = \frac{1}{2} \omega_{ab} \omega^{ab}$, the *vorticity propagation equation*,

$$h^a_b \dot{\omega}^b + \frac{2}{3} \theta \omega^a = \sigma^a{}_b \omega^b + \frac{1}{2} \eta^{abcd} u_b \dot{u}_{c;d}, \qquad (4.52)$$

where $\omega^a = \frac{1}{2} \eta^{abcd} u_b \omega_{cd}$ is the vorticity vector, the *shear propagation equation*,

$$h^f_a h^g_b \dot{\sigma}_{fg} + \frac{2}{3} \theta \sigma_{ab} = h^f_a h^g_b \dot{u}_{(f;g)} + \dot{u}_a \dot{u}_b - \omega_a \omega_b - \sigma_{af} \sigma^f{}_b$$

$$+ h_{ab} \left(\frac{1}{3} \omega^2 + \frac{2}{3} \sigma^2 - \frac{1}{3} \dot{u}^c{}_{;c} \right)$$

$$+ \frac{1}{2} \pi_{ab} - E_{ab}, \qquad (4.53)$$

the so-called $(0,\nu)$ field equation [see Ellis (1971) for the explanation of this terminology],

$$\frac{2}{3} h^a_b \theta^{,b} - h^a_b \sigma^{bc}{}_{;d} h^d_c - \eta^{acdf} u_c (\omega_{d;f} + 2\omega_d \dot{u}_f) = q^a, \qquad (4.54)$$

the *divergence of vorticity equation*,

$$h^b_a \omega^a{}_{;b} = \dot{u}_a \omega^a, \qquad (4.55)$$

and the *magnetic part of the Weyl tensor*,

$$H_{ab} = 2 \dot{u}_{(a} \omega_{b)} - h^t_a h^s_b (\omega_{(t}{}^{g;c} + \sigma_{(t}{}^{g;c}) \eta_{s)fgc} u^f. \qquad (4.56)$$

The projections mentioned above of the equations of motion and the energy conservation equation yield the *equations of motion of matter*,

$$(\mu + p) \dot{u}^a = -h^{ab}(p_{,b} + \pi^c{}_{b;c} + \dot{q}_b) - (\omega^{ab} + \sigma^{ab} + \frac{4}{3} \theta h^{ab}) q_b, \qquad (4.57)$$

and the *energy conservation equation*,

$$\dot{\mu} + \theta(\mu + p) + \pi_{ab} \sigma^{ab} + q^a{}_{;a} + \dot{u}^a q_a = 0. \qquad (4.58)$$

Finally the projections mentioned above of the Bianchi identities yield equations involving E_{ab} and H_{ab} which are roughly analogous to Maxwell's equations in electromagnetic theory. The equations which emerge from the projections of (4.49) are the *div-E equation*,

$$h_g^b\, E^{gd}{}_{;f}\, h_d^f = -3\omega^f\, H_f^b + \eta^{bapq}\, u_a\, \sigma^d{}_p\, H_{qd} + \frac{1}{3} h_f^b\, \mu^{;f}$$

$$+ \frac{1}{2}\left\{ -\pi^{bd}{}_{;d} + u^b\, \sigma_{cd}\, \pi^{cd} - 3\omega^{bd}\, q_d + \sigma^{bd}\, q_d \right.$$

$$\left. - \frac{2}{3}\theta\, q^b + \pi^{bd}\, \dot{u}_d \right\}, \tag{4.59}$$

the *div-H* equation,

$$h_g^b\, H^{gd}{}_{;f}\, h_d^f = 3\omega^f\, E_f^b - \eta^{bapq}\, u_a\, \sigma^d{}_p\, E_{qd} + (\mu + p)\,\omega^b$$

$$+ \frac{1}{2}\eta^b{}_{qac}\, u^q\, q^{a;c} + \frac{1}{2}\eta^b{}_{qac}\, u^q\, (\omega^{dc} + \sigma^{dc})\, \pi^a{}_d, \tag{4.60}$$

the *Ė* equation,

$$h_f^b\, \dot{E}^{fg}\, h_g^t + \theta\, E^{bt} = -h_a^{(b}\, \eta^{t)rsd}\, u_r\, H^a{}_{s;d} + 2\, H^{(b}{}_s\, \eta^{t)drs}\, u_d\, \dot{u}_r + E^{(t}{}_s\, \omega^{b)s}$$

$$+ 3\, E^{(t}{}_s\, \sigma^{b)s} - h^{tb}\, E^{dp}\, \sigma_{dp} - \frac{1}{2}(\mu + p)\,\sigma^{tb}$$

$$- \frac{1}{6} h^{tb}\,\{\dot{\mu} + \theta\,(\mu + p)\} - q^{(b}\, \dot{u}^{t)} - \frac{1}{2} u^{(b}\, \dot{q}^{t)}$$

$$- \frac{1}{2} q^{t;b} + \frac{1}{2}\{\omega^{c(b} + \sigma^{c(b}\}\, u^{t)}\, q_c + \frac{1}{6}\theta\, u^{(t}\, q^{b)}$$

$$- \frac{1}{2}\dot{\pi}^{bt} + \pi^{c(b}\, u^{t)}\, \dot{u}_c - \frac{1}{2}\{\omega^{c(b} + \sigma^{c(b}\}\, \pi^{t)}{}_c$$

$$- \frac{1}{6}\theta\, \pi^{bt}, \tag{4.61}$$

and the *Ḣ* equation,

$$h_f^b\, \dot{H}^{fg}\, h_g^t + \theta\, H^{bt} = h_a^{(b}\, \eta^{t)rsd}\, u_r\, E^a{}_{s;d} - 2\, E^{(b}{}_s\, \eta^{t)drs}\, u_d\, \dot{u}_r + H^{(t}{}_s\, \omega^{b)s}$$

$$+ 3\, H^{(t}{}_s\, \sigma^{b)s} - h^{tb}\, H^{dp}\, \sigma_{dp} - q^{(t}\, \omega^{b)}$$

$$- \frac{1}{2}\eta^{(t}{}_{rad}\,\{\omega^{b)d} + \sigma^{b)d}\}\, u^r\, q^a$$

$$- \frac{1}{2}\eta^{(b}{}_{rad}\, \pi^{t)a;d}\, u_r + \frac{1}{2}\eta^{(b}{}_{rad}\, u^{t)}\, u^r\,\{\omega^{cd} + \sigma^{cd}\}\, \pi^a{}_c. \tag{4.62}$$

We shall henceforth assume that the space–time is a perturbed Robertson–Walker space–time. In this sense we have a background (unperturbed) space–time which is isotropic with respect to every integral curve of the time-like vector field u^a. The metric tensor g_{ab} of this background space–time is the Robertson–Walker metric [given by (4.102) below]. The background energy–momentum–stress tensor is given by (4.35) specialized to a perfect fluid (with $q^a = 0 = \pi^{ab}$ and the scalars μ, p satisfy $h_b^a\, \mu_{,a} = 0 = h_b^a\, p_{,a}$), which is the most general allowable matter distribution consistent with isotropy. The background Weyl tensor vanishes on account of isotropy and so the

64 Inhomogeneous aspects of cosmology

background space–time is conformally flat. The integral curves of the vector field u^a must be geodesic, vorticity-free, shear-free, and with the scalar θ satisfying $h_b^a\,\theta_{,a} = 0$. Hence in this background (4.42) specializes to

$$u_{a;b} = \frac{1}{3}\,\theta\,h_{ab}. \tag{4.63}$$

The Ellis–Bruni (1989) approach to describing perturbations of this isotropic background is to work with gauge-invariant small quantities rather than with small perturbations of the background metric tensor. Such gauge-invariant variables are identified as being those variables which vanish in the background space–time. Hence for an isotropic background the small, of first-order, gauge-invariant Ellis–Bruni variables are $E_{ab}, H_{ab}, \dot{u}^a, \omega_{ab}$ (or equivalently the vorticity vector ω^a), $\sigma_{ab}, X_b = h_b^a\,\mu_{,a}, Y_b = h_b^a\,p_{,a}, Z_b = h_b^a\,\theta_{,a}, \pi^{ab}$, and q^a. The equations to be satisfied by these small quantities are obtained from (4.51)–(4.62) neglecting nonlinear terms involving these small quantities with the proviso that Raychaudhuri's equation (4.51) and the energy conservation equation (4.58) need to be modified to be expressed in terms of the gauge-invariant variables. This latter is achieved by taking the partial derivatives of (4.51) and (4.58) and then using the projection tensor h_b^a to project the differentiated equations orthogonal to u^a while at the same time neglecting second-order small quantities. The result, starting with (4.51), is the equation

$$\dot{Z}^a + \theta\,Z^a = \dot{\theta}\,\dot{u}^a + h^{ab}\,(\dot{u}^c{}_{;c})_{,b} - \frac{1}{2}X^a - \frac{3}{2}Y^a, \tag{4.64}$$

while this sequence of operations applied to (4.58) yields

$$\dot{X}^a + \frac{4}{3}X^a = \dot{\mu}\,\dot{u}^a - h^{ab}\,(q^c{}_{;c})_{,b} - (\mu + p)\,Z^a - \theta\,Y^a. \tag{4.65}$$

The background value of $\dot{\theta}$ to be substituted into (4.64) is obtained from Raychaudhuri's equation (4.51) specialized to the isotropic background to read

$$\dot{\theta} = -\frac{1}{3}\theta^2 - \frac{1}{2}(\mu + 3p). \tag{4.66}$$

The background value of $\dot{\mu}$ to be substituted into (4.65) is obtained from (4.58) specialized to the isotropic background to read

$$\dot{\mu} = -\theta\,(\mu + p). \tag{4.67}$$

All equations are tensorial with no particular coordinate system specified and so this Ellis–Bruni perturbation theory is both gauge invariant and covariant.

4.3 Gravitational waves

For perturbations of isotropic cosmological models which describe pure gravitational radiation only the perturbed shear σ_{ab} of the matter world lines and the perturbed anisotropic stress π_{ab} of the matter distribution are required along with the derived perturbation variables E_{ab} and H_{ab}. These represent, in the present context, so-called tensor perturbations in a language often used in perturbation theory. All of the remaining Ellis–Bruni gauge-invariant variables will be assumed to vanish from now on. From the Ricci identities linearized in terms of the small non-vanishing gauge-invariant variables we have the equations:

$$\dot{\sigma}_{ab} + \frac{2}{3}\theta\sigma_{ab} - \frac{1}{2}\pi_{ab} + E_{ab} = 0, \qquad (4.68)$$

$$\sigma^{ab}{}_{;b} = 0, \qquad (4.69)$$

and

$$H_{ab} = -\sigma_{(a}{}^{g;c}\eta_{b)fgc}\,u^f. \qquad (4.70)$$

The equations of motion of matter provide us with

$$\pi^{ab}{}_{;b} = 0, \qquad (4.71)$$

while the Bianchi identities yield the equations:

$$E^{ab}{}_{;b} + \frac{1}{2}\pi^{ab}{}_{;b} = 0, \qquad (4.72)$$

and

$$H^{ab}{}_{;b} = 0, \qquad (4.73)$$

from (4.59) and (4.60) and

$$\dot{E}^{ab} + \theta E^{ab} = -\eta^{(a}{}_{rsd}H^{b)s;d}\,u^r - \frac{1}{2}\sigma^{ab} - \frac{1}{2}\dot{\pi}^{ab} - \frac{1}{6}\theta\pi^{ab}, \qquad (4.74)$$

$$\dot{H}^{ab} + \theta H^{ab} = \eta^{(a}{}_{rsd}E^{b)s;d}\,u^r - \frac{1}{2}\eta^{(a}{}_{rsd}\pi^{b)s;d}\,u^r. \qquad (4.75)$$

For solutions of (4.68)–(4.75) which describe simple gravitational waves, by which we mean gravitational waves with clearly identifiable wavefronts, we need to identify in the Robertson–Walker background space–time a family of null hypersurfaces which will play the role of the histories of the wavefronts. One way of achieving this is to assume that the basic gauge-invariant variables σ_{ab} and π_{ab} have an arbitrary dependence on a scalar function $\phi(x^a)$ (say). The derived gauge-invariant variables E_{ab} and H_{ab} will then inherit this dependence. Thus we write

$$\sigma_{ab} = s_{ab}\,F(\phi), \quad \pi_{ab} = \Pi_{ab}\,F(\phi), \qquad (4.76)$$

where F is an *arbitrary* function of $\phi(x^a)$ and s_{ab}, Π_{ab} are symmetric, trace-free, and orthogonal to u^a with respect to the background metric g_{ab}. Gauge-invariant variables of this form were first introduced into the Ellis–Bruni formalism by Hogan and Ellis (1997) and the idea was developed by Hogan and O'Shea (2002a, 2002b) and O'Shea (2004a, 2004b). The introduction of arbitrary functions into solutions of Einstein's equations describing gravitational waves goes back to the pioneering work of Trautman (1962). Such waves have been described by Trautman as carrying arbitrary information. Using (4.76) we obtain from (4.68) the equations

$$E_{ab} = \left(\frac{1}{2}\Pi_{ab} + p_{ab}\right)F + m_{ab}\,F', \qquad (4.77)$$

where $F' = dF/d\phi$ and

$$p_{ab} = -\dot{s}_{ab} - \frac{2}{3}\theta\, s_{ab}, \tag{4.78}$$

$$m_{ab} = -\dot{\phi}\, s_{ab}, \tag{4.79}$$

where $\dot{\phi} = \phi_{,a} u^a \neq 0$. From (4.69) we have

$$s^{ab}{}_{;b} = 0 \quad \text{and} \quad s^{ab}\phi_{,b} = 0. \tag{4.80}$$

Using (4.70) we find that

$$H_{ab} = q_{ab} F + l_{ab} F', \tag{4.81}$$

with

$$q_{ab} = -s_{(a}{}^{g;c} \eta_{b)fgc}\, u^f, \tag{4.82}$$

$$l_{ab} = -s_{(a}{}^{g} \eta_{b)fgc}\, u^f\, \phi^{,c}. \tag{4.83}$$

By (4.71) we have

$$\Pi^{ab}{}_{;b} = 0 \quad \text{and} \quad \Pi^{ab}\phi_{,b} = 0. \tag{4.84}$$

Next (4.72) and (4.73) yield the two sets of equations:

$$p^{ab}{}_{;b} = 0, \quad p^{ab}\phi_{,b} + m^{ab}{}_{;b} = 0, \quad m^{ab}\phi_{,b} = 0, \tag{4.85}$$

and

$$q^{ab}{}_{;b} = 0, \quad q^{ab}\phi_{,b} + l^{ab}{}_{;b} = 0, \quad l^{ab}\phi_{,b} = 0, \tag{4.86}$$

respectively. Finally (4.74) and (4.75) yield the two sets of equations:

$$\dot{\Pi}^{ab} + \frac{2}{3}\theta\,\Pi^{ab} = -\dot{p}^{ab} - \theta\, p^{ab} - u_r\, q^{(a}{}_{s;d}\, \eta^{b)rsd} - \frac{1}{2}(\mu + p)\, s^{ab}, \tag{4.87}$$

$$\dot{\phi}\,\Pi^{ab} = -\dot{\phi}\,p^{ab} - \dot{m}^{ab} - \theta\, m^{ab} - u_r\left(q^{(a}{}_{s}\phi_{,d} + l^{(a}{}_{s;d}\right)\eta^{b)rsd}, \tag{4.88}$$

$$\dot{\phi}\, m^{ab} + l^{(a}{}_{s}\,\eta^{b)rsd}\, u_r\, \phi_{,d} = 0, \tag{4.89}$$

and

$$\dot{q}^{ab} - u_r\, p^{(a}{}_{s;d}\,\eta^{b)rsd} + \theta\, q^{ab} = 0, \tag{4.90}$$

$$\dot{\phi}\, q^{ab} + \dot{l}^{ab} + \theta\, l^{ab} - u_r\left(p^{(a}{}_{s}\phi_{,d} + m^{(a}{}_{s;d}\right)\eta^{b)rsd} = 0, \tag{4.91}$$

$$\dot{\phi}\, l^{ab} - m^{(a}{}_{s}\,\eta^{b)rsd}\, u_r\, \phi_{,d} = 0, \tag{4.92}$$

respectively.
Putting

$$V^{ab} = m^{ab} + i l^{ab}, \tag{4.93}$$

we can combine (4.89) and (4.92) into the single complex equation

$$\dot{\phi} V^{ab} = i V^{(a}{}_s \eta^{b)rsd} u_r \phi_{,d}. \tag{4.94}$$

We calculate from this that

$$\dot{\phi} \eta_{bpql} V^{ab} u^p = 2 i V^a{}_{[q} \lambda_{l]}, \tag{4.95}$$

where $\lambda_a = h^b_a \phi_{,b}$. When this is substituted into the right-hand side of (4.94) we obtain

$$\dot{\phi} V^{ab} = \dot{\phi} V^{ab} + \dot{\phi}^{-1} \phi_{,d} \phi^{,d} V^{ab}. \tag{4.96}$$

With $\dot{\phi} \neq 0$ and $V^{ab} \neq 0$ we must have

$$\phi_{,d} \phi^{,d} = 0. \tag{4.97}$$

Thus the hypersurfaces $\phi(x^a) = $ constant in the background Robertson–Walker space–time must be null.

Substituting into (4.88) for p^{ab} given by (4.78) and for m^{ab} given by (4.79) results in

$$\dot{\phi} \Pi^{ab} = 2 \dot{\phi} \dot{s}^{ab} + \frac{5}{3} \theta \dot{\phi} s^{ab} + \ddot{\phi} s^{ab} - u_r \left(q^{(a}{}_s \phi_{,d} + l^{(a}{}_{s;d} \right) \eta^{b)rsd}. \tag{4.98}$$

Substituting q^{ab} and l^{ab} into this from (4.82) and (4.83), respectively, finally results in

$$s'_{ab} + \left(\frac{1}{2} \phi^{,d}{}_{;d} - \frac{1}{3} \theta \dot{\phi} \right) s_{ab} = -\frac{1}{2} \dot{\phi} \Pi_{ab}, \tag{4.99}$$

where $s'_{ab} = s_{ab;d} \phi^{,d}$. This is a *propagation equation for* s_{ab} along the null geodesics tangent to $\phi_{,a}$ in the background Robertson–Walker space–time.

Consider now (4.87). Substituting into this for p^{ab} from (4.78) and for q^{ab} from (4.82) and using the first of (4.70) and the Ricci identities satisfied by s^{ab} we arrive at a *wave equation for* s^{ab}:

$$s^{ab;d}{}_{;d} - \frac{2}{3} \theta \dot{s}^{ab} - \left(\frac{1}{3} \dot{\theta} + \frac{4}{9} \theta^2 \right) s^{ab} + \left(p - \frac{1}{3} \mu \right) s^{ab} = -\dot{\Pi}^{ab} - \frac{2}{3} \theta \Pi^{ab}. \tag{4.100}$$

The key equations satisfied by $\phi(x^a)$, s^{ab}, and Π^{ab} are (4.80), (4.84), (4.97), (4.99), and (4.100). A lengthy calculation, indicated in some detail in Hogan and O'Shea (2002a), establishes that these equations are mathematically consistent with each other and also that the remaining equations in (4.85), (4.86), (4.90), and (4.91) are automatically satisfied. We now seek to exhibit solutions of the key equations having the property that the corresponding perturbed Weyl tensor components given by (4.77) and (4.81) are type N in the Petrov classification with degenerate principal null direction $\phi_{,a}$ and thus with

$$E_{ab} \phi^{,b} = 0 = H_{ab} \phi^{,b}. \tag{4.101}$$

We can then interpret physically the perturbations of the isotropic cosmological models as describing pure gravitational radiation having the null hypersurfaces $\phi = $ constant, in the background Robertson–Walker (RW) space–times, as the histories of the wavefronts. To achieve this we first set out to discover some naturally occurring null hypersurfaces in the RW space–times. We begin with the general RW line-element in standard form:

$$ds^2 = R^2(t) \frac{\{(dx^1)^2 + (dx^2)^2 + (dx^3)^2\}}{\left(1 + \frac{k}{4} r^2\right)^2} - dt^2, \qquad (4.102)$$

with $R(t)$ the scale factor, $r^2 = (x^1)^2 + (x^2)^2 + (x^3)^2$, and $k = 0, \pm 1$ the Gaussian curvature of the space-like hypersurfaces $t =$ constant. We can transform (4.102) into the following forms suitable for our purposes (Hogan, 1988):

$$ds^2 = R^2(t) \{dx^2 + p_0^{-2} f^2 (dy^2 + dz^2)\} - dt^2, \qquad (4.103)$$

with $p_0 = 1 + (K/4)(y^2 + z^2)$, $K =$ constant, and $f = f(x)$ and (i) if $k = +1$ then $K = +1$ and $f(x) = \sin x$; (ii) if $k = 0$ then $K = 0, +1$ with $f(x) = 1$ when $K = 0$ and $f(x) = x$ when $K = +1$; (iii) if $k = -1$ then $K = 0, \pm 1$ with $f(x) = (1/2) e^x$ when $K = 0$, $f(x) = \sinh x$ when $K = +1$ and $f(x) = \cosh x$ when $K = -1$.

Case (i) above arises because when $k = +1$ the closed universe model with line-element (4.102) has $t =$ constant hypersurfaces having induced line-element $dl^2 = R^2(t) ds_0^2$ with

$$ds_0^2 = dx^2 + \sin^2 x \, (d\vartheta^2 + \sin^2 \vartheta \, d\varphi^2), \qquad (4.104)$$

and y, z are stereographic coordinates given by $y + iz = 2 e^{i\varphi} \cot(\vartheta/2)$.

Case (ii) occurs because when $k = 0$ the open, spatially flat universe model with line-element (4.102) has $t =$ constant hypersurfaces having induced line-element $dl^2 = R^2(t) ds_0^2$ with

$$ds_0^2 = dx^2 + dy^2 + dz^2, \qquad (4.105)$$

or

$$ds_0^2 = dx^2 + x^2 (d\vartheta^2 + \sin^2 \vartheta \, d\varphi^2), \qquad (4.106)$$

and in the latter we introduce the stereographic coordinates y, z again as above.

Case (iii) corresponding to (4.102) with $k = -1$ has three possibilities. The $t =$ constant hypersurfaces can, modulo the factor $R^2(t)$, be viewed as the future sheet of a unit time-like hypersphere $\mathcal{H}^3$ in 4-dimensional Minkowskian space–time $\mathcal{M}^4$. Writing the line-element of $\mathcal{M}^4$ as

$$ds_0^2 = (dz^1)^2 + (dz^2)^2 + (dz^3)^2 - (dz^4)^2, \qquad (4.107)$$

the hypersphere $\mathcal{H}^3$ has equation

$$(z^1)^2 + (z^2)^2 + (z^3)^2 - (z^4)^2 = -1, \quad z^4 > 0. \qquad (4.108)$$

The three possibilities of case (iii) are due to the different ways in which one can parametrize (4.108). The first possibility is to take

$$z^1 + iz^2 = (y + iz) p_0^{-1} \sinh x, \qquad (4.109)$$

$$z^3 = \left(\frac{1}{4}(y^2 + z^2) - 1\right) p_0^{-1} \sinh x, \qquad (4.110)$$

$$z^4 = \cosh x, \qquad (4.111)$$

with $p_0 = 1 + \frac{1}{4}(y^2 + z^2)$. Substitution into (4.107) yields
$$ds_0^2 = dx^2 + p_0^{-2} \sinh^2 x \, (dy^2 + dz^2). \tag{4.112}$$

The next possibility is to take
$$z^1 + iz^2 = \frac{1}{2} e^x (y + iz), \tag{4.113}$$
$$z^3 = \frac{1}{4} e^x (y^2 + z^2 - 1) + e^{-x}, \tag{4.114}$$
$$z^4 = \frac{1}{4} e^x (y^2 + z^2 + 1) + e^{-x}, \tag{4.115}$$

and now (4.107) becomes
$$ds_0^2 = dx^2 + \frac{1}{4} e^{2x} (dy^2 + dz^2). \tag{4.116}$$

Finally we have
$$z^1 + iz^2 = (y + iz) p_0^{-1} \cosh x, \tag{4.117}$$
$$z^3 = \sinh x, \tag{4.118}$$
$$z^4 = \left(\frac{1}{4}(y^2 + z^2) + 1\right) p_0^{-1} \cosh x, \tag{4.119}$$

with $p_0 = 1 - \frac{1}{4}(y^2 + z^2)$. With this (4.107) takes the form
$$ds_0^2 = dx^2 + p_0^{-2} \cosh^2 x \, (dy^2 + dz^2). \tag{4.120}$$

In the space–times with line-elements (4.103) the hypersurfaces
$$\phi(x^a) = x - T(t) = \text{constant}, \tag{4.121}$$

with $dT/dt = R^{-1}(t)$ are *null hypersurfaces*. These null hypersurfaces are generated by null geodesics having expansion
$$\frac{1}{2} \phi^{;a}{}_{;a} = \frac{f'}{R^2 f} + \frac{\dot{R}}{R^2}. \tag{4.122}$$

Here $f' = df/dx$ and $\dot{R} = dR/dt$. The integral curves of the vector field $\partial/\partial t$ are the world lines of the fluid particles. The components of this vector field are u^a and using (4.121) we find that
$$2\phi_{,a;b} = \xi_a \phi_{,b} + \xi_b \phi_{,a} + \phi_{,d}{}^{;d} g_{ab}, \tag{4.123}$$

with
$$\xi_a = -\frac{f'}{f} \phi_{,a} + R \phi_{,d}{}^{;d} u_a. \tag{4.124}$$

It follows from (4.123) that the null geodesic integral curves of the vector field $\phi_{,a}$ are *shear-free* in the optical sense (Robinson and Trautman, 1983).

For convenience we have used the same coordinate labels x, y, z, t in all of the special cases above. The ranges of the coordinates will of course be different in the different cases. For example in case (ii) $x \in (-\infty, +\infty)$ if $K = 0$ whereas $x \in [0, +\infty)$ if $K = +1$. In addition the shear-free null hypersurfaces (4.121) differ from case to case, and also within cases (ii) and (iii), as can be appreciated by considering the intersections of these null hypersurfaces with the space-like hypersurface $t = $ constant. In case (i) the intersection is a 2-sphere. In case (ii) the intersection is a 2-sphere if $K = +1$ and a 2-plane if $K = 0$. Hence (4.121) describes two quite different families of shear-free null hypersurfaces that can arise in an open, spatially flat universe. In case (iii) the intersection of (4.121) with $t = $ constant can be a 2-space of constant positive curvature (if $K = +1$), of constant negative curvature (if $K = -1$), or of zero curvature (if $K = 0$), yielding three different families of shear-free null hypersurfaces in a $k = -1$ open universe. A geometrical origin of these subcases is described in Hogan (1988).

To solve the key equations (4.80), (4.84), (4.97), (4.99), and (4.100) satisfied by $\phi(x^a)$, s^{ab}, and Π^{ab}, beginning with (4.121), we note that since s^{ab} and Π^{ab} are each trace-free and orthogonal to u_a and to $\phi_{,a}$ they each have only two independent components. If we label the coordinates in (4.102) by $x^1 = x$, $x^2 = y$, $x^3 = z$, $x^4 = t$ then we begin by writing the non-vanishing components of s^{ab} and Π^{ab} as $s^{33} = -s^{22} = \alpha(x, y, z, t)$ and $s^{23} = s^{32} = \beta(x, y, z, t)$ and $\Pi^{33} = -\Pi^{22} = A(x, y, z, t)$ and $\Pi^{23} = \Pi^{32} = B(x, y, z, t)$. It is convenient to define a null tetrad via the 1-forms:

$$m_a \, dx^a = \frac{1}{\sqrt{2}} R \, p_0^{-1} f \, (dy + i dz), \qquad (4.125)$$

$$\bar{m}_a \, dx^a = \frac{1}{\sqrt{2}} R \, p_0^{-1} f \, (dy - i dz), \qquad (4.126)$$

$$k_a \, dx^a = R \, dx - dt, \qquad (4.127)$$

$$l_a \, dx^a = -\frac{1}{2} (R \, dx + dt). \qquad (4.128)$$

Thus m^a, $\bar{m}^a$, k^a, l^a, with the bar denoting complex conjugation, constitute a tetrad of null vectors with all scalar products among them vanishing except for $m_a \bar{m}^a = 1$ and $k_a l^a = -1$. In terms of this tetrad we have

$$s^{ab} = \bar{s} \, m^a \, m^b + s \, \bar{m}^a \, \bar{m}^b, \qquad (4.129)$$

with

$$\bar{s} = -R^2 \, p_0^{-2} f^2 \, (\alpha + i\beta), \qquad (4.130)$$

and

$$\Pi^{ab} = \bar{\Pi} \, m^a \, m^b + \Pi \, \bar{m}^a \, \bar{m}^b, \qquad (4.131)$$

with

$$\bar{\Pi} = -R^2 \, p_0^{-2} f^2 \, (A + iB). \qquad (4.132)$$

Substitution of (4.129) into the first of (4.80) reveals that the real-valued functions α and β must satisfy the Cauchy–Riemann equations:

$$\frac{\partial}{\partial y}(p_0^{-4} \alpha) - \frac{\partial}{\partial z}(p_0^{-4} \beta) = 0, \qquad (4.133)$$

$$\frac{\partial}{\partial y}(p_0^{-4}\,\beta) + \frac{\partial}{\partial z}(p_0^{-4}\,\alpha) = 0. \tag{4.134}$$

Similarly (4.131) substituted into the first of (4.84) results in the real-valued functions A and B satisfying the same equations:

$$\frac{\partial}{\partial y}(p_0^{-4}\,A) - \frac{\partial}{\partial z}(p_0^{-4}\,B) = 0, \tag{4.135}$$

$$\frac{\partial}{\partial y}(p_0^{-4}\,B) + \frac{\partial}{\partial z}(p_0^{-4}\,A) = 0. \tag{4.136}$$

To work with (4.133) and (4.134) it is convenient to make the change of dependent variables from α, β to α_0, β_0 with

$$\alpha_0 = f^3\,R^3\,\alpha \quad \text{and} \quad \beta_0 = f^3\,R^3\,\beta, \tag{4.137}$$

which clearly satisfy (4.133) and (4.134) which can be written in the form

$$\frac{\partial}{\partial \bar{\zeta}}\{p_0^{-4}(\alpha_0 + i\beta_0)\} = 0, \tag{4.138}$$

with $\zeta = y + iz$. Hence

$$\alpha_0 + i\beta_0 = p_0^4\,\mathcal{G}(\zeta, x, t), \tag{4.139}$$

where $\mathcal{G}$ is an analytic function of ζ. Now (4.130) becomes

$$\bar{s} = -R^{-1}\,p_0^2\,f^{-1}\,\mathcal{G}(\zeta, x, t). \tag{4.140}$$

The propagation equation (4.99) for s^{ab} along the integral curves of $\phi^{\cdot a}$ gives A, B in terms of α_0, β_0. Then using (4.132) the end result can be written in the form

$$\bar{\Pi} = -2\,R^{-2}\,p_0^2\,f^{-1}(D\mathcal{G} + \dot{R}\,\mathcal{G}), \tag{4.141}$$

with the operator D defined by

$$D = \frac{\partial}{\partial x} + R\,\frac{\partial}{\partial t} = \frac{\partial}{\partial x} + \frac{\partial}{\partial T}, \tag{4.142}$$

and $T(t)$ introduced in (4.121). It follows in particular from (4.132) and (4.141) that now $A + iB$ is analytic in ζ and so the Cauchy–Riemann equations (4.135) and (4.136) are automatically satisfied. With s^{ab} now given by (4.129) and (4.140) and with Π^{ab} given by (4.131) and (4.141) we substitute these results into the wave equation (4.100). The result, after a lengthy calculation, is the remarkably simple equation:

$$D^2\mathcal{G} + k\,\mathcal{G} = 0, \tag{4.143}$$

with $k = 0, \pm 1$ labelling the RW backgrounds with line-elements (4.102). In deriving (4.143) we have made use of the fact that $f(x)$ in (4.103), and the cases (i)–(iii), satisfies

$$f'' = -k\,f \quad \text{and} \quad (f')^2 + k\,f^2 = K. \tag{4.144}$$

The solutions of (4.143) are easily seen to be given by

$$\mathcal{G}(\zeta, x, t) = a_1(x + T) + a_2, \tag{4.145}$$

for $k = 0$, by

$$\mathcal{G}(\zeta, x, t) = a_1 \sin \frac{(x + T)}{2} + a_2 \cos \frac{(x + T)}{2}, \tag{4.146}$$

for $k = +1$, and by

$$\mathcal{G}(\zeta, x, t) = a_1 \sinh \frac{(x + T)}{2} + a_2 \cosh \frac{(x + T)}{2}, \tag{4.147}$$

for $k = -1$. In all cases $a_1(\zeta, x - T)$ and $a_2(\zeta, x - T)$ are arbitrary functions of their arguments. Corresponding to these solutions the Weyl tensor components (4.77) and (4.81) can be written as

$$E^{ab} + iH^{ab} = -2 R^{-2} p_0^2 f^{-1} \frac{\partial}{\partial x}(\mathcal{G} F) m^a m^b, \tag{4.148}$$

where now $F = F(\phi) = F(x - T)$ is the arbitrary function introduced at the outset. It is clear from (4.148) that now the equations in (4.101) are satisfied.

The propagation equation (4.99) for s^{ab} along the integral curves of the null vector field $\phi^{\cdot a}$ demonstrates that if $s^{ab} = 0$ then $\Pi^{ab} = 0$. For the gravitational wave perturbations we have calculated above there is an interesting converse property: *if* $\Pi^{ab} = 0$ *then* $s^{ab} = 0$ *provided* $\mu + p \neq 0$. To see this we first note from (4.141) that $\Pi^{ab} = 0$ implies that

$$D\mathcal{G} + \dot{R}\mathcal{G} = 0. \tag{4.149}$$

Substituting this into the wave equation (4.143) results in

$$(\dot{R}^2 - R\ddot{R} + k)\mathcal{G} = 0. \tag{4.150}$$

Using Eistein's field equations for the background RW space–time, the fluid proper density μ and the isotropic pressure p satisfy

$$p = -\frac{\dot{R}^2}{R^2} - \frac{2\ddot{R}}{R} - \frac{k}{R}, \tag{4.151}$$

$$\mu = \frac{3\dot{R}^2}{R^2} + \frac{3k}{R^2}, \tag{4.152}$$

from which we obtain

$$\frac{2}{R^2}(\dot{R}^2 - R\ddot{R} + k) = \mu + p, \tag{4.153}$$

and so (4.150) can be written simply as

$$(\mu + p)\mathcal{G} = 0, \tag{4.154}$$

from which the above result follows.

4.4 Cosmic background radiation

Electromagnetic test fields on the RW space–times are described by a Maxwell tensor with components $F_{ab} = -F_{ba}$ satisfying Maxwell's source-free field equations on the RW space–times. With respect to the distinguished time-like direction u^a in these space–times we can define the electric and magnetic parts of this Maxwell field by

$$E_a = F_{ab}\, u^b \quad \text{and} \quad H_a = {}^*F_{ab}\, u^a, \tag{4.155}$$

respectively, with the components of the dual of the Maxwell tensor given by ${}^*F_{ab} = \frac{1}{2}\eta_{abcd}\, F^{cd}$. Maxwell's source-free equations on the RW space–times read (Ellis, 1971):

$$E^a{}_{;a} = 0 \quad H^a{}_{;a} = 0, \tag{4.156}$$

and

$$\dot{E}^a + \frac{2}{3}\theta E^{ab} = -\eta^{abcd}\, u_b\, H_{e;d}, \tag{4.157}$$

$$\dot{H}^a + \frac{2}{3}\theta H^a = \eta^{abcd}\, u_b\, E_{e;d}. \tag{4.158}$$

To solve these we introduce a 4-potential σ^a satisfying (Ellis and Hogan, 1997)

$$\sigma^a\, u_a = 0 \quad \text{and} \quad \sigma^a{}_{;a} = 0, \tag{4.159}$$

from which the Maxwell tensor F_{ab} is calculated using

$$F_{ab} = \sigma_{b;a} - \sigma_{a;b}. \tag{4.160}$$

Now (4.156) and (4.158) are satisfied and (4.157) simplifies to the wave equation

$$\sigma^{a;b}{}_{;b} = \frac{1}{2}(\mu - p)\sigma^a. \tag{4.161}$$

In parallel with (4.76) we look for solutions of (4.159) and (4.161) of the form

$$\sigma^a = s^a\, F(\phi) \quad \text{with} \quad s^a\, u_a = 0. \tag{4.162}$$

Now, taking advantage of the arbitrariness of the function F as before, the second of (4.159) yields

$$s^a{}_{;a} = 0 \quad \text{and} \quad s^a\, \phi_{,a} = 0, \tag{4.163}$$

while the wave equation (4.161) with $s^a \neq 0$ provides us with the equations:

$$\phi^{,a}\, \phi_{,a} = 0, \tag{4.164}$$

$$s'_a + \frac{1}{2}\phi^b{}_{;b}\, s_a = 0, \tag{4.165}$$

$$s^{a;b}{}_{;b} = \frac{1}{2}(\mu - p)\, s^a, \tag{4.166}$$

where $s'_a = s_{a;b}\, \phi^{,b}$.

We begin with some simple sinusoidal, monochromatic electromagnetic waves on a $k = 0$ RW space–time (Hogan and O'Farrell, 2011). The line-element of the space–time is given by (4.102) with $k = 0$. First take

$$\phi = \delta_{\alpha\beta}\, n^\alpha\, x^\beta - T, \tag{4.167}$$

with Greek indices taking values 1, 2, 3. Also $\delta_{\alpha\beta}$ is the 3-dimensional Kronecker delta and we see from (4.102) with $k = 0$ that $\delta_{\alpha\beta} = R^{-2} g_{\alpha\beta}$. As before $T(t)$ satisfies $dT/dt = R^{-1}$. The constants n^α satisfy

$$\mathbf{n} \cdot \mathbf{n} = \delta_{\alpha\beta} n^\alpha n^\beta = 1. \tag{4.168}$$

It is somewhat superfluous to write the Kronecker delta in, for example, (4.167) when we could simply write $\phi = n^\alpha x^\alpha - T$. However if we differentiate (4.167) with respect to x^λ we obtain $\phi_{,\lambda} = \delta_{\alpha\lambda} n^\alpha$ and some people find it confusing to write this as $\phi_{,\lambda} = n^\lambda$ since the indices on either side of the equation are in different positions. With the choice (4.167) and the metric given via the line-element (4.102) with $k = 0$ we see that (4.164) is satisfied. As candidate solutions of (4.159) and (4.161) we consider the potential 1-form

$$\sigma_a \, dx^a = \lambda \delta_{\alpha\beta} (B^\beta \cos\phi + C^\beta \sin\phi). \tag{4.169}$$

Here λ is a real constant, and $\mathbf{B} = (B^\beta)$ and $\mathbf{C} = (C^\beta)$ are constant 3-vectors related to the 3-vector $\mathbf{n} = (n^\beta)$ by

$$\mathbf{B} = \mathbf{b} \times \mathbf{n} \quad \text{and} \quad \mathbf{C} = \mathbf{b} - (\mathbf{b} \cdot \mathbf{n}) \mathbf{n}, \tag{4.170}$$

for any 3-vector $\mathbf{b}$ such that $\mathbf{b} \cdot \mathbf{b} = 1$. We have the following useful relations:

$$\mathbf{n} \cdot \mathbf{B} = \mathbf{n} \cdot \mathbf{C} = \mathbf{B} \cdot \mathbf{C} = 0, \qquad \mathbf{B} \cdot \mathbf{B} = \mathbf{C} \cdot \mathbf{C} = 1 - (\mathbf{b} \cdot \mathbf{n})^2, \tag{4.171}$$

and

$$\delta_{\mu\alpha} \delta_{\nu\beta} \left\{ n^\alpha n^\beta + (\mathbf{B} \cdot \mathbf{B})^{-1} (B^\alpha B^\beta + C^\alpha C^\beta) \right\} = \delta_{\mu\nu}. \tag{4.172}$$

Both terms in (4.169) are of the form (4.162) with s_a taking the form

$$s_a = (\delta_{\alpha\beta} a^\beta, 0) \quad \text{with} \quad \mathbf{a} \cdot \mathbf{n} = 0, \tag{4.173}$$

and with $\mathbf{a}$ a constant 3-vector. It is now straightforward to confirm that (4.163)–(4.166) are satisfied. The expansion of these waves is given by (4.122) with $f = 1$ and so we see that the wavefronts are expanding on account of the expansion of the universe. The unit 3-vector $\mathbf{n}$ gives the direction of propagation of the waves at any point in the 3-space $t = $ constant. With (4.169) in (4.160) and (4.155) we find that $E_a = (E_\alpha, 0), H_a = (H_\alpha, 0)$ and

$$E_\alpha = \frac{\lambda}{R} \delta_{\alpha\beta} (-B^\beta \sin\phi + C^\beta \cos\phi), \tag{4.174}$$

$$H_\alpha = -\frac{\lambda}{R} \delta_{\alpha\beta} (B^\beta \cos\phi + C^\beta \sin\phi). \tag{4.175}$$

Also

$$F_{\alpha 4} = -F_{4\alpha} = E_\alpha \quad \text{and} \quad F_{\alpha\beta} = -F_{\beta\alpha} = R \, \epsilon_{\alpha\beta\gamma} H_\gamma, \tag{4.176}$$

where $\epsilon_{\alpha\beta\gamma}$ is the 3-dimensional Levi-Cività permutation symbol. The electromagnetic energy–momentum tensor associated with these waves has components

$$M_{ab} = \frac{1}{2} (F_{ac} F_b{}^c + {}^*F_{ac} {}^*F_b{}^c). \tag{4.177}$$

Here $M_{ab} = M_{ba}$ and $M^a{}_a = 0$. Explicit calculation of these components for the electromagnetic waves above, making use of (4.172) in particular to simplify $M_{\mu\nu}$, yields

$$M_{\mu\nu} = \frac{\lambda^2}{R^2} (\mathbf{B} \cdot \mathbf{B}) \, \delta_{\mu\alpha} \, \delta_{\nu\beta} \, n^\alpha \, n^\beta, \tag{4.178}$$

$$M_{\mu 4} = -\frac{\lambda^2}{R^3} (\mathbf{B} \cdot \mathbf{B}) \, \delta_{\mu\alpha} \, n^\alpha, \tag{4.179}$$

$$M_{44} = \frac{\lambda^2}{R^4} (\mathbf{B} \cdot \mathbf{B}). \tag{4.180}$$

These equations can be written succinctly as

$$M_{ab} = \frac{\lambda^2}{R^2} (\mathbf{B} \cdot \mathbf{B}) \, \phi_{,a} \, \phi_{,b}. \tag{4.181}$$

We note that $\mathbf{B} \cdot \mathbf{B}$ is written in terms of $\mathbf{b}$ and $\mathbf{n}$ in (4.171). To obtain a model of the background electromagnetic radiation in the $k = 0$ RW universe we sum the electromagnetic energy tensors for all such wave systems labelled by the directions of $\mathbf{b}$ and $\mathbf{n}$. Summing (which involves integrating) M_{ab} over the directions of $\mathbf{b}$ results in

$$M_{\mu\nu} = \frac{8\pi\lambda^2}{3R^2} \delta_{\mu\alpha} \, \delta_{\nu\beta} \, n^\alpha \, n^\beta, \tag{4.182}$$

$$M_{\mu 4} = -\frac{8\pi\lambda^2}{3R^3} \delta_{\mu\alpha} \, n^\alpha, \tag{4.183}$$

$$M_{44} = \frac{8\pi\lambda^2}{3R^4}. \tag{4.184}$$

Finally summing (i.e. integrating) over the directions of $\mathbf{n}$ we arrive at the energy–momentum–stress tensor of the background electromagnetic radiation (remembering that we are neglecting anisotropies) with components $\mathcal{M}_{ab}$ given by

$$\mathcal{M}_{\mu\nu} = \frac{c_0^2}{R^2} \delta_{\mu\nu}, \tag{4.185}$$

$$\mathcal{M}_{\mu 4} = 0, \tag{4.186}$$

$$\mathcal{M}_{44} = \frac{3 c_0^2}{R^4}, \tag{4.187}$$

with $c_0^2 = 32\pi^2\lambda^2/9$. This can be summarized as

$$\mathcal{M}_{ab} = \mu_e \, u_a \, u_b + p_e \, g_{ab}, \tag{4.188}$$

with

$$p_e = \frac{1}{3}\mu_e = \frac{c_0^2}{R^4}. \tag{4.189}$$

The subscripts on the proper density and isotropic pressure of this perfect fluid energy–momentum–stress tensor are chosen to reflect their origin in the electromagnetic waves. The fluid 4-velocity is u^a and the equation of state of the fluid is (4.189). We note that $\mathcal{M}^{ab}{}_{;b} = 0$, as it should be.

76 Inhomogeneous aspects of cosmology

The construction of an isotropic model of background radiation above can be extended to include the RW universes with $k = \pm 1$ in (4.102). The end result will again be the energy–momentum–stress tensor (4.188). Anticipating this we can simplify the derivation by calculating this tensor along any one of the integral curves of the vector field u^a and then extending the result to all curves on the basis of isotropy. For convenience we shall work on the curve $r = 0$, with r defined following (4.102). With s_a given by (4.173) we now have $s^a = (f^2 \, a^\alpha / R^2, 0)$. Also ϕ given by (4.167) satisfies (4.164) with now g_{ab} given by (4.102). Calculating $s^a{}_{;a}$ along $r = 0$ we find that $s^a{}_{;a} = 0$ and $(s^a{}_{;a})_{,b}\, u^b = 0$ but $(s^a{}_{;a})_{,\alpha} \neq 0$. This latter is given along $r = 0$ by

$$(s^a{}_{;a})_{,\alpha} = -\frac{k}{2\,R^2}\, s_\alpha. \tag{4.190}$$

Now (4.163)–(4.165) are satisfied along $r = 0$. The derivation of the wave equation (4.166) has made use of the first of (4.163). If we do not assume that the first of (4.163) is satisfied everywhere then the wave equation (4.166) is modified to read

$$s_{a;d}{}^{;d} = \frac{1}{2}\, (\mu - p)\, s_a + h_a^b \, (s^d{}_{;d})_{,b}, \tag{4.191}$$

along $r = 0$. This is automatically satisfied when $a = 4$. With $s_\alpha = \delta_{\alpha\beta}\, a^\beta$ we find that along $r = 0$

$$s_{\alpha;d}{}^{;d} = -\left(\frac{\ddot R}{R} + \frac{2\,\dot R^2}{R^2} + \frac{3\,k}{2\,R^2}\right) s_\alpha = \frac{1}{2}\,(\mu - p)\, s_\alpha - \frac{k}{2\,R^2}\, s_\alpha. \tag{4.192}$$

We have used (4.151) and (4.152) to arrive at the final equality here. Hence (4.191) is satisfied along $r = 0$. Since the passage from s_a to F_{ab} involves only first derivatives of s_a, we obtain for F_{ab}, and thus for M_{ab} along $r = 0$, for the cases $k = \pm 1$, the same expressions as those given by (4.174)–(4.176) and (4.181) in the case $k = 0$. Hence summing the electromagnetic energy tensors with respect to the directions of the 3-vectors **b** and **n** produces again the isotropic energy–momentum–stress tensor (4.188).

A simple family of gravitational waves in a $k = 0$ RW universe, analogous to the electromagnetic waves described by (4.16), are given by the special case of the waves described in Section 4.3 with

$$\sigma_{ab}\, dx^a\, dx^b = \lambda\, R\, \delta_{\alpha\mu}\, \delta_{\beta\nu} \Big\{(B^\mu B^\nu - C^\mu C^\nu)\cos\phi$$

$$+ (B^\mu C^\nu + C^\mu B^\nu)\sin\phi\Big\} dx^\alpha\, dx^\beta, \tag{4.193}$$

and

$$\pi_{ab} = \frac{2\,\dot R}{R}\, \sigma_{ab}, \tag{4.194}$$

with λ, B^α, C^α, and ϕ as in (4.16). Now the only non-vanishing components of the electric and magnetic parts of the perturbed Weyl tensor are

$$E_{\alpha\beta} = \lambda \, \delta_{\alpha\mu} \, \delta_{\beta\nu} \{(C^\mu \, C^\nu - B^\mu \, B^\nu) \sin \phi + (B^\mu \, C^\nu + C^\mu \, B^\nu) \cos \phi\}, \tag{4.195}$$

$$H_{\alpha\beta} = \lambda \, \delta_{\alpha\mu} \, \delta_{\beta\nu} \{-(B^\mu \, C^\nu + C^\mu \, B^\nu) \sin \phi + (C^\mu \, C^\nu - B^\mu \, B^\nu) \cos \phi\}. \tag{4.196}$$

Equivalently the non-vanishing components of the perturbed Weyl tensor are given by

$$C_{\alpha\beta\gamma\delta} = -R^2 \, \epsilon_{\alpha\beta\lambda} \, \epsilon_{\gamma\delta\sigma} \, E_{\lambda\sigma}, \quad C_{\alpha\beta\gamma 4} = R \, \epsilon_{\alpha\beta\sigma} \, H_{\sigma\gamma}, \quad C_{\alpha 4 \beta 4} = E_{\alpha\beta}. \tag{4.197}$$

The analogue of the electromagnetic energy tensor (4.177) is the Bel–Robinson tensor (Penrose and Rindler, 1984)

$$M_{abcd} = \frac{1}{4} \left(C_a{}^p{}_b{}^q \, C_{cpdq} + {}^* C_a{}^p{}_b{}^q {}^* C_{cpdq} \right). \tag{4.198}$$

This satisfies $M_{(abcd)} = M_{abcd}$ (the round brackets denoting symmetrization) and $M^a{}_{acd} = 0$. For the current gravitational wave perturbations (4.195) and (4.196) we find that M_{abcd} can be simplified to read

$$M_{abcd} = 2 \, \lambda^2 \, (\mathbf{B} \cdot \mathbf{B})^2 \, \phi_{,a} \, \phi_{,b} \, \phi_{,c} \, \phi_{,d}, \tag{4.199}$$

with $\mathbf{B} \cdot \mathbf{B}$ given by (4.171). Summing this, by integration, over all possible directions of $\mathbf{b}$ gives

$$M_{abcd} = \frac{64 \, \pi \, \lambda^2}{15} \, \phi_{,a} \, \phi_{,b} \, \phi_{,c} \, \phi_{,d}, \tag{4.200}$$

and then summing this, by integration, over all possible directions of $\mathbf{n}$ we arrive at an isotropic tensor $\mathcal{M}_{abcd}$ associated with the gravitational radiation background given by

$$\mathcal{M}_{abcd} = \frac{c_1^2}{R^4} \left\{ u_a \, u_b \, u_c \, u_d + 2 \, h_{(ab} \, u_c \, u_{d)} + \frac{1}{5} \, h_{(ab} \, h_{cd)} \right\}, \tag{4.201}$$

with $c_1^2 = 256 \, \pi^2 \, \lambda^2 / 15$. We have derived this expression starting with gravity wave perturbations of an RW universe with line-element (4.102) with $k = 0$. Its extension to include the cases $k = \pm 1$ follows the pattern of the extension to these cases of the electromagnetic example above. The details can be found in Hogan and O'Farrell (2011).

Since $\mathcal{M}_{abcd}$ is constructed from the Bel–Robinson tensor, which is quadratic in the perturbed Weyl tensor components, it is a second-order, gauge-invariant, small quantity. It has dimensions of (length)$^{-4}$ and its divergence is given by

$$\mathcal{M}^{abcd}{}_{;d} = \frac{2 \, c_1^2}{3 \, R^4} \, \theta \left\{ u^a \, u^b \, u^c + \frac{1}{3} \, (h^{ab} \, u^c + h^{bc} \, u^a + h^{ac} \, u^b) \right\}, \tag{4.202}$$

after making use of (4.63). Alternatively we have

$$\left(\mathcal{M}^{abcd} \, u_a \, u_b \right)_{;d} = 0, \tag{4.203}$$

where
$$\mathcal{M}^{abcd} u_a u_b = \frac{c_1^2}{R^4}\left(u_c u_d + \frac{1}{3} h_{cd}\right). \tag{4.204}$$

This tensor is symmetric, trace-free, and divergence-free. These are properties we associate with an energy–momentum–stress tensor. On the other hand, (4.204) has dimensions of (length)$^{-4}$. Using the sale factor $R(t)$ we can define an energy–momentum–stress tensor with dimensions of (length)$^{-2}$ by multiplying (4.204) by R_0^2 for some $R_0 = R(t_0) \neq 0$. This results in the energy–momentum–stress tensor

$$T^{ab} = \mu_g\, u^a u^b + p_g\, h^{ab}, \tag{4.205}$$

with

$$p_g = \frac{1}{3}\mu_g = \frac{c_1^2\, R_0^2}{3\, R^4}. \tag{4.206}$$

The subscripts here on the proper density and isotropic pressure reflect the gravitational wave origin of these variables. This is one way in which an energy–momentum–stress tensor can be associated with a cosmic background of gravitational radiation. It would be interesting to incorporate this second-order perturbation of the RW energy–momentum–stress tensor into the Ellis–Bruni perturbation theory as a source of second-order perturbations of isotropic cosmologies. Anisotropies in this gravitational radiation background will thus be of third order.

5
Black holes

In comparison with the Newtonian theory of gravitation Einstein's theory of gravitation, general relativity, predicts two important new phenomena: gravitational waves and black holes. Chapters 2 and 4 of this book have been mainly devoted to gravitational waves. Black holes will now be the subject of this chapter and the next chapter. Following a review of the basic properties of black holes some selected topics dealing with classical and quantum aspects of black holes are presented. In particular we consider the formation of trapped surfaces during gravitational collapse, the scattering properties of a high-speed Kerr black hole, the spontaneous creation of de Sitter universes inside a black hole, and the effect of metric fluctuations on Hawking radiation. Higher dimensional black holes will be studied in the next chapter.

5.1 Introduction: Basic properties of black holes

We begin with a summary of some of the basic properties of black holes which will be useful in this chapter. More information on black holes can be found for instance in Frolov and Zelnikov (2011), Frolov and Novikov (1998), and Poisson (2004) and references therein. We emphasize that the notation and sign conventions we use continue to be those summarized in Appendix A and we use units in which the speed of light *in vacuo* is $c = 1$ and the gravitational constant is $G = 1$. Uniqueness theorems on the existence of black holes ensure that in vacuum, any static and asymptotically flat solution of the Einstein's field equations which is spherically symmmetric must coincide with the Schwarzschild solution. A similar result holds for stationary, vacuum, asymptotically flat space–times with the Kerr black hole as the unique black hole in this case. These theorems have been extended to electrically charged black holes resulting in the Reissner–Nordström black hole in the static case and Kerr–Newman black hole for the stationary case. In the latter two cases the combined set of Einstein-Maxwell vacuum field equations have to be solved. Only electrically neutral black holes will be considered here however.

The Schwarzschild solution describes the gravitational field outside an isolated, static, and spherically symmetric body with mass M. In terms of the Schwarzschild coordinates $x^i = (r, \theta, \phi, t)$ the metric tensor of the space–time is given via the line-element

$$ds^2 = -f(r)\,dt^2 + f^{-1}(r)\,dr^2 + r^2\,d\Omega^2, \tag{5.1}$$

where $d\Omega^2 = d\theta^2 + \sin^2\theta \, d\phi^2$ is the line-element of the unit 2-dimensional sphere and the function f is given by

$$f(r) = 1 - \frac{r_H}{r}, \tag{5.2}$$

with $r_H \equiv 2M$. The line-element (5.1) is singular at $r = r_H$. However since the curvature invariant, $R^{ijkl}R_{ijkl} = 48M/r^6$, remains finite at $r = r_H$ we see that $r = r_H$ is not a true singularity of the space–time but is in fact a *coordinate* singularity which can be eliminated by a suitable change of coordinates. On the other hand, $r = 0$ is a true singularity of the space–time. Various alternative forms of the Schwarzschild line-element are made available by introducing different local coordinates. For example, advanced and retarded time coordinates v and u, respectively, are defined by

$$dv = dt + \frac{dr}{f(r)}, \qquad du = dt - \frac{dr}{f(r)}, \tag{5.3}$$

and (5.2) becomes

$$ds^2 = -f(r) \, du \, dv + r^2 \, d\Omega^2. \tag{5.4}$$

From (5.3) and the definition of $f(r)$ we see that in the domain $r > r_H$ any future-directed radial trajectory of a particle travelling with the velocity of light is outgoing (ingoing) and corresponds to $u = $ constant ($v = $ constant). The radial coordinate r in (5.4) is now a function of the pair of null coordinates (u,v). Introducing the so-called tortoise radial coordinate r_* defined by

$$r_* = \int \frac{dr}{f(r)} = r + 2M \log\left|\frac{r}{r_H} - 1\right|, \tag{5.5}$$

one obtains the following implicit form of $r(u,v)$

$$v - u = 2r_*. \tag{5.6}$$

It follows from (5.5) that r_* is well defined for $r > r_H$ and for $r < r_H$ and vanishes at the singularity $r = 0$. The coordinate singularity $r = r_H$ is still present in the form (5.4) of the line-element. It can be eliminated if the null coordinates (u, v) are replaced by the Kruskal null coordinates (U, V) via the coordinate transformations

$$U = -e^{-\frac{u}{2r_H}}, \qquad V = e^{\frac{v}{2r_H}}, \tag{5.7}$$

With this new system of coordinates the line-element (5.4) is transformed into

$$ds^2 = -\frac{32M^3}{r} e^{-\frac{r}{r_H}} dU \, dV + r^2 \, d\Omega^2, \tag{5.8}$$

showing that the metric tensor is now regular at $r = r_H$. The implicit relation $r = r(U,V)$ is now given by

$$-UV = \left(\frac{r}{r_H} - 1\right) e^{\frac{r}{r_H}}. \tag{5.9}$$

An interesting property of the Kruskal coordinates is that by allowing U and V to take positive and negative values (so that $U = \mp e^{-\frac{u}{2r_H}}$ and $V = \pm e^{\frac{v}{2r_H}}$) one obtains

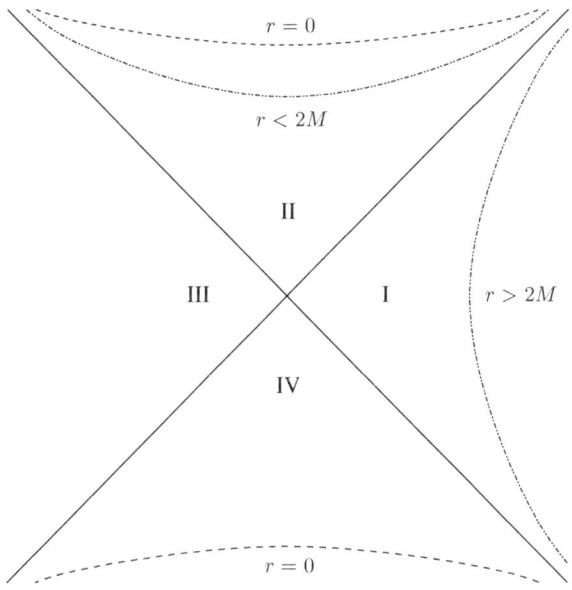

Fig. 5.1 The Kruskal diagram.

an extension of the space–time manifold. This extension is represented in Figure 5.1 where the angular coordinates (θ, ϕ) are suppressed. In Figure 5.1 only the domains I and II are physically relevant. The curves $r = \text{constant}$ correspond to the hyperbolae $UV = \text{constant}$, with the two particular values $U = 0$ and $V = 0$ corresponding to $r = r_H$. Light rays are straight lines parallel to the coordinate axes $U = 0$ and $V = 0$. In domain I any future-directed radial light ray $U = \text{constant}$ is outgoing and reaches infinity, while any future-directed radial light ray $V = \text{constant}$ is ingoing and crosses $r = r_H$. In domain II both radial light rays $U = \text{contant}$ and $V = \text{constant}$ reach the singularity $r = 0$. Therefore domain II has no causal influence on domain I and for this reason is called the black hole interior. The semi-axis $V > 0$, or $r = r_H$, separates domains I and II and is called the future event horizon of the black hole. Regions III and IV result from the analytic extension of the Schwarzschild solution and are not physically relevant. For instance in a gravitational collapse they have to be replaced by the interior solution of the Einstein field equations applicable to the collapsing body.

The event horizon $r = r_H$ is a stationary null hypersurface. Its normal $n_i = \partial_i r$ is a null vector and thus $n \cdot n|_{r=r_H} = g^{rr}|_{r=r_H} = f(r_H) = 0$, with the dot denoting, for convenience, the scalar product with respect to the space–time metric. Since the Schwarzschild space–time is static the vector field $\xi_{(t)} = \partial/\partial t$ is a Killing vector field which is also null on the horizon in the sense that $\xi_{(t)} \cdot \xi_{(t)}|_{r_H} = g_{tt}|_{r_H} = -f(r_H) = 0$. Thus the hypersurface $r = r_H$ is also a Killing horizon.

A compactified form of the Kruskal diagram is given by the Carter–Penrose diagram in Figure 5.2. The significance of the symbols in the diagram are: $\mathcal{J}^+$ is future null infinity ($v = +\infty$, u finite); $\mathcal{J}^-$ is past null infinity ($u = -\infty$, v finite); i^0 is spatial

82 Black holes

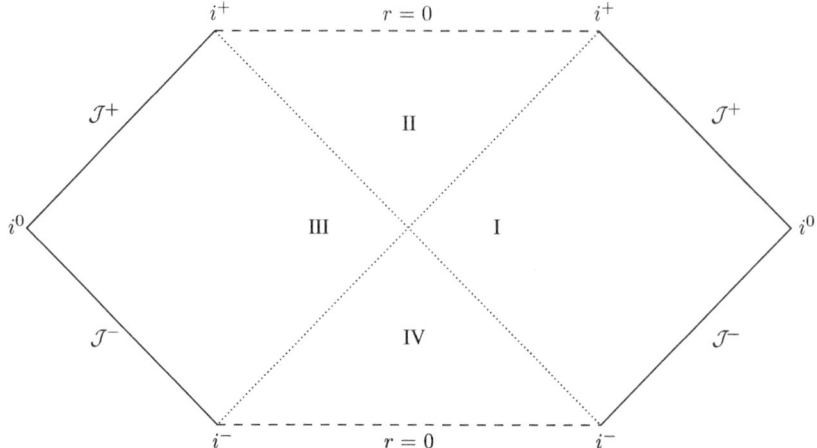

Fig. 5.2 The Carter–Penrose diagram for the Schwarzschild solution; dotted lines represent the event horizons $r = 2M$.

infinity ($r = +\infty$, t finite); i^+ is future time-like infinity ($t = +\infty$, r finite); and i^- is past time-like infinity ($t = -\infty$, r finite). A local coordinate system which is often convenient and which is intermediate between the Schwarzschild coordinates (t, r) and the double null coordinates (u, v) is provided by the Eddington–Finkelstein coordinates (v, r) or (u, r). Using the definitions (5.3) one immediately obtains

$$ds^2 = dv\,(2\,dr - f(r)\,dv) + r^2\,d\Omega^2, \tag{5.10}$$

or

$$ds^2 = -du\,(2\,dr + f(r)\,du) + r^2\,d\Omega^2. \tag{5.11}$$

We now consider the rotating Kerr black hole. Its line-element expressed in so-called Boyer–Lindquist coordinates $x^i = (r, \theta, \phi, t)$ takes the form

$$ds^2 = -\left(1 - \frac{2Mr}{\rho^2}\right) dt^2 + \frac{\rho^2}{\Delta} dr^2 + \rho^2\,d\theta^2 + \frac{\Sigma \sin^2 \theta}{\rho^2} d\phi^2 - \frac{4Mar\sin^2\theta}{\rho^2} dt\,d\phi, \tag{5.12}$$

where

$$\rho^2 = r^2 + a^2 \cos^2 \theta, \tag{5.13}$$
$$\Delta = r^2 - 2Mr + a^2, \tag{5.14}$$
$$\Sigma = (r^2 + a^2)^2 + a^2 \Delta \sin^2 \theta. \tag{5.15}$$

The mass of the black hole is M, the angular momentum is $J = Ma$, and a is called the angular momentum parameter (the angular momentum per unit mass or specific angular momentum). The line-element (5.12) is singular at the values of r

Introduction: Basic properties of black holes   83

for which $\Delta = 0$ or $\rho^2 = 0$. However the curvature invariant $R^{ijkl}R_{ijkl}$ for the Kerr solution is

$$R^{ijkl}R_{ijkl} = \frac{48M^2(r^2 - a^2\cos^2\theta)(\rho^4 - 16a^2r^2\cos^2\theta)}{\rho^{12}}, \qquad (5.16)$$

which shows that only $\rho^2 = 0$ is a true singularity of the space–time. The values of r for which $\Delta = 0$ correspond to coordinate singularities. Using Kerr–Schild coordinates (described below) one shows that the singularity corresponding to $\rho^2 = 0$ is located on the ring with equation

$$x^2 + y^2 = a^2, \qquad z = 0. \qquad (5.17)$$

The event horizon can be found by looking for a null hypersurface with equation $r = $ constant and thus by solving the equation

$$g^{ij}\partial_i r\,\partial_j r = g^{rr} = \frac{\Delta}{\rho^2} = 0. \qquad (5.18)$$

This equation possesses the two real solutions or roots:

$$r = r_\pm = M \pm \sqrt{M^2 - a^2}, \qquad (5.19)$$

whenever $a < M$. Only the largest root $r = r_+$ corresponds to the event horizon and the domain $r < r_+$ is the interior of the Kerr black hole. The smaller root $r = r_-$ is called the inner apparent horizon and is also a Cauchy horizon which is a null hypersurface beyond which predictability breaks down as it lies outside the domain of dependence of any spatial slice covering all space. When $a = M$ the Kerr black hole is said to be extremal and when $a > M$ no real solution of (5.18) exists and in this case the Kerr solution is the space–time model of the gravitational field of a naked singularity rather than a black hole. The Carter–Penrose diagram of the Kerr black hole is shown in Figure 5.3.

The Kerr solution is stationary and admits the two Killing vector fields $\xi_{(t)} = \partial/\partial t$ and $\xi_{(\phi)} = \partial/\partial \phi$. We recall that a Killing vector field generates a one-parameter group of isometries of space–time. A necessary and sufficient condition that ξ be a Killing vector field is that it satisfies

$$\xi_{i;j} + \xi_{j;i} = 0, \qquad (5.20)$$

with covariant differentiation, as always, denoted by a semicolon. An important property of a Killing vector field is that, given a geodesic γ with tangent u, the scalar quantity $\xi \cdot u$ is conserved along γ and is called a constant of the motion. In addition a Killing vector field satisfies $\xi_{i;jk} = R_{ijkl}\xi^l$ on account of (5.20) and the Ricci identities. In particular for an observer with 4-velocity $u^i = (0, 0, \dot\phi, \dot t)$ the scalar product $u \cdot \xi_{(\phi)}$ represents an angular momentum and it is a conserved quantity. Whenever it vanishes the observer has no angular momentum and is referred to as a zero angular momentum observer (ZAMO). In this case

$$\Omega \equiv \frac{d\phi}{dt} = -\frac{g_{t\phi}}{g_{\phi\phi}} = \frac{2Mar}{\Sigma}. \qquad (5.21)$$

84 Black holes

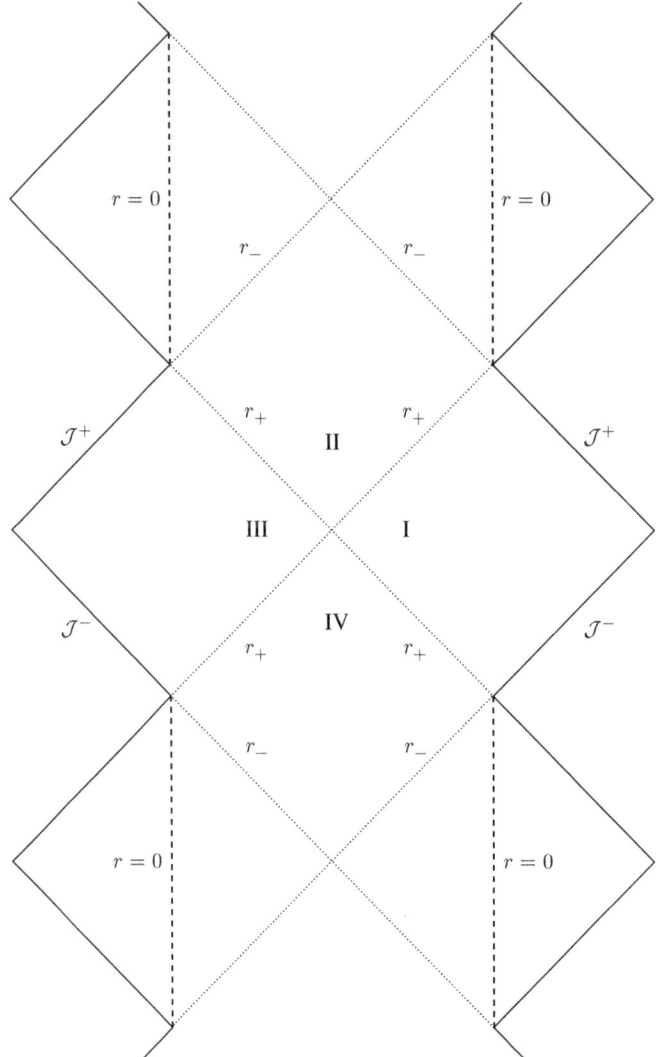

Fig. 5.3 The Carter–Penrose diagram of the Kerr black hole with $a < M$; dotted lines represent the event horizon $r = r_+$ and the inner apparent horizon $r = r_-$.

Although the observer has no angular momentum, Ω, which actually represents an angular velocity, is not zero. This phenomenon is known as the dragging of inertial frames. At the event horizon $r = r_+$ the quantity Ω is given by

$$\Omega_H = \frac{a}{a^2 + r_+^2}. \tag{5.22}$$

This is called the angular velocity of the black hole. It can be shown that the vector

$$\eta \equiv \xi_{(t)} + \Omega_H \, \xi_{(\phi)}, \tag{5.23}$$

is a Killing vector and that it is null at the event horizon and thus $\eta \cdot \eta|_H = 0$. Therefore the event horizon is also a Killing horizon.

The Kerr metric posseses the following Killing tensor of order 2:

$$K_{ij} = \Delta k_{(i} k_{j)} + r^2 g_{ij}, \tag{5.24}$$

where k_i and l_i are the null vectors defined by (suppressing indices for convenience)

$$k = \frac{r^2 + a^2}{\Delta} \xi_{(t)} - \frac{\partial}{\partial r} + \frac{a}{\Delta} \xi_{(\phi)}, \tag{5.25}$$

$$l = \frac{r^2 + a^2}{\Delta} \xi_{(t)} + \frac{\partial}{\partial r} + \frac{a}{\Delta} \xi_{(\phi)}, \tag{5.26}$$

and the round brackets on the indices in (5.24) denote symmetrization. $K_{ij} u^i u^j$ is conserved along a geodesic with tangent u^i, and it is related to the Carter constant $Q = K_{ij} u^i u^j - (u \cdot \xi_{(\phi)} + a u \cdot \xi_{(t)})^2$. In general a Killing tensor of order m with coordinate components $K_{i_1..i_m}$ is defined to be totally symmetric and to obey the condition

$$K_{(j;i_1...i_m)} = 0. \tag{5.27}$$

Killing tensors do not arise in any natural way from groups of isometries as is the case with Killing vectors. However they do give rise to constants of the motion of the form $K_{i_1...i_m} u^{i_1} \ldots u^{i_m}$.

We end this brief review of some of the basic properties of black holes by presenting the Kerr–Schild form of the black hole metric (for more details see Section 5.3). This form corresponds to writing the Kerr line-element as

$$ds^2 = \eta_{ij} dx^i dx^j + 2H k_i k_j dx^i dx^j, \tag{5.28}$$

where η_{ij} are the components of the Minkowski metric tensor given by the line-element

$$\eta_{ij} dx^i dx^j = dx^2 + dy^2 + dz^2 - dt^2, \tag{5.29}$$

k_i is the null covariant vector field defined by the 1-form

$$k_i dx^i = \frac{rx + ay}{r^2 + a^2} dx + \frac{ry - ax}{r^2 + a^2} dy + \frac{z}{r} dz - dt, \tag{5.30}$$

and

$$H = \frac{Mr^3}{r^4 + a^2 z^2}. \tag{5.31}$$

In addition we have $k^i = g^{ij} k_j = \eta^{ij} k_j$ and the components of the inverse of the metric tensor are given by $g^{ij} = \eta^{ij} - H k^i k^j$. The system of coordinates $x^i = (x, y, z, t)$ used in (5.28) corresponds to asymptotically rectangular Cartesian coordinates and time with (x, y, z) satisfying

$$\frac{x^2 + y^2}{r^2 + a^2} + \frac{z^2}{r^2} = 1. \tag{5.32}$$

We note that the time coordinate t here is not the same as in (5.12). The Kerr–Schild form of the Schwarzschild solution is obtained by putting $a = 0$ in the above equations.

5.2 Collapsing null shells and trapped surface formation

A lapse of about forty years passed between the discovery of the Schwarzschild solution in 1916 and its interpretation as a black hole, with the latter name first used by J. A. Wheeler in a public lecture in 1967. The renewal of interest in this subject has been in large part due to the theorems on singularities by Penrose (1965) establishing the conditions for the inevitability of their formation in gravitational collapse, and by Hawking (1967) concerning the initial singularity in cosmology [for a comprehensive discussion see Hawking and Ellis (1973) and Frolov and Novikov (1998)]. These results required the notions of trapped surfaces and horizons which are basic ingredients of the general properties of a black hole.

The presence of an event horizon depends on the causal structure of the space–time manifold $\mathcal{M}$. Let $J^+(p)$ be the region of $\mathcal{M}$ which can be connected to some event p by any unbroken future-directed, time-like or light-like curve. The region $J^+(p)$ is called the causal future of p. One easily extends this definition to the causal past $J^-(p)$ of p and to the causal future/past $J^\pm(S)$ of some set of events S. For instance, the region $J^-(\mathcal{J}^+)$ contains all the null geodesics which reach future null infinity $\mathcal{J}^+$. In contrast the domain $\mathcal{B}$ of the space–time manifold $\mathcal{M}$ defined by

$$\mathcal{B} \equiv \mathcal{M} - J^-(\mathcal{J}^+), \tag{5.33}$$

is causally disconnected to null infinity. If $\mathcal{B}$ is not an empty set a black hole exists and $\mathcal{B}$ is called the interior of the black hole (e.g. region II in Figures 5.1 or 5.2 in the Schwarzschild example). Its boundary $H = \partial \mathcal{B}$ is the event horizon of the black hole. The event horizon is a null hypersurface. It has a teleological nature in the sense that one needs to know the whole history of the black hole in order to obtain it. This remark is of particular importance for black holes which are dynamically formed after, say, successive accretions of matter.

The dynamical aspect of black hole formation is encapsulated in the notion of a trapped surface. Let us consider a compact 2-dimensional space-like submanifold S and both sets of ingoing and outgoing null geodesics orthogonal to S. If the expansion rate of the ingoing and outgoing geodesics is everywhere negative on S then S is said to be a trapped surface. It is said to be marginally trapped if one of the two null geodesic congruences is expansion-free. If during the formation of the black hole one visualizes a 3-dimensional space-like surface Σ then the boundary of the marginally trapped region within Σ is called the apparent horizon. The apparent horizon evolves with time and eventually ends in the formation of the event horizon. It is locally given by the set of events where future-directed null geodesics have zero expansion. According to the theorem by Hawking (1971) concerning the non-decreasing property of the surface area of a black hole, the apparent horizon is always contained within the event horizon. The Schwarzschild and Kerr black holes are stationary and describe eternal black holes. The Schwarzshild apparent horizon and event horizon coincide while for the Kerr solution $r = r_+$ is both an event horizon and an outer apparent horizon and $r = r_-$ is an inner apparent horizon and also a Cauchy horizon.

Trapped surfaces play an important role in the theorems on singularities. In 1970 Penrose and Hawking proved a theorem stating that if, during a gravitational collapse, a trapped surface is formed and some additional conditions are satisfied (in particular energy conditions on the matter content), then a singularity necessarily forms. A singularity represents a pathological behaviour of space–time and generally corresponds to the divergence of curvature invariants. It may also lead to a breakdown of causality as for instance in the case of what is called a naked singularity (see for example the Kerr solution). The singularity which arises in a generic gravitational collapse is admissible if it is hidden to external observers by an event horizon. This idea was first articulated by Penrose in his famous cosmic censorship hypothesis. A weak 1969 version of the latter reads: 'singularities formed in a generic gravitational collapse cannot causally influence events near $\mathcal{J}^+$'. On the other hand, a strong 1971 version reads: 'singularities formed in a generic gravitational collapse must be space-like' or 'no singularity formed in a gravitational collapse can be seen by an observer unless he falls into it'. No demonstrations of any of the versions of this hypothesis have yet been given and no physically relevant counter-examples have been found.

In an attempt to build counter-examples to cosmic censorship Penrose (1973) considered a thin shell collapsing from infinity with the speed of light, hereafter referred to as a null shell. If M is the total gravitational mass of the shell and if an apparent horizon with area $\mathcal{A}$ forms during the collapse then the so-called Gibbons–Penrose isoperimetric inequality [see also Gibbons (1972)]

$$\mathcal{A} \leq 4\pi \, (2M)^2, \tag{5.34}$$

holds. The validity of (5.34) was originally proved for closed convex shells and was later extended to the non-convex case by Gibbons (1997) who also concluded that it is impossible to construct a contradiction to cosmic censorship using collapsing shells. Thorne proposed an alternative conjecture, the hoop conjecture, according to which horizons form when and only when a mass M gets compacted in every direction into a region whose circumference $\mathcal{C}$ in every direction satisfies

$$\mathcal{C} \leq 4\pi M. \tag{5.35}$$

The two inequalities encapsulate the idea that horizons form whenever matter is sufficiently compacted in a region of space. They do not refer to singularities but refer to trapped surfaces and apparent horizons (the nature of the horizon is not specified in the hoop conjecture), and a mathematical justification of their validity is easier to propose because of the local nature of these surfaces. Let us consider the Penrose (1973) model of a closed and convex thin shell $\mathcal{S}$ of dust collapsing from infinity with the speed of light in an initially Minkowskian space–time. The space–time history of $\mathcal{S}$ is a singular null hypersurface $\mathcal{N}$ whose interior remains flat. Let n be a future-directed null vector tangent to the null generators of $\mathcal{N}$ and let k be a future-directed null vector tangent to the congruence of outgoing null geodesics orthogonal to the shell. Then $n \cdot n = 0 = k \cdot k$ and we take $n \cdot k = -1$ as they are both future-directed. The surface stress–energy tensor of the null shell is $T_\mathcal{S}^{\alpha\beta} = \mu \, n^\alpha n^\beta \, \delta(\Phi)$ where μ is the surface energy density of the shell, $\Phi(x) = 0$ is the equation of $\mathcal{N}$, and $\delta(\Phi)$ the Dirac delta function which is singular on $\mathcal{N}$. The expansion θ, the shear σ, and the twist ω of the outgoing null congruence are discontinuous across $\mathcal{N}$. Let us now assume that at some moment of time an outer apparent horizon forms and coincides with $\mathcal{S}$. Then the outer value of the expansion vanishes, and if one integrates Raychaudhuri's equation for the congruence tangent to k,

$$\frac{d\theta}{d\lambda} + \frac{\theta^2}{2} + \sigma^2 + \omega^2 = R_{ij} k^i k^j, \tag{5.36}$$

one gets

$$\theta = 8\pi\mu. \tag{5.37}$$

The extrinsic curvature K at any point of $\mathcal{S}$ calculated in Euclidean 3-space is thus $K = 2\theta = 16\pi\mu$. On the other hand, we have the Minkowski (1903) inequality of classical differential geometry

$$16\pi\mathcal{A} \leq \left(\int K \, dS \right)^2, \tag{5.38}$$

which holds for any convex closed surface of area $\mathcal{A}$. Combining (5.37) and (5.38) one immediately obtains the Gibbons–Penrose isoperimetric inequality (5.34). The gravitational mass of the shell $M = \int \mu \, dS$ is conserved during the collapse and it is equal to the Bondi advanced mass, the ADM mass, and the Hawking quasi-local mass. Had a model violating this inequality been constructible then no event horizon could ever have developed in the collapse, fundamentally contradicting cosmic censorship. Furthermore if cosmic censorship holds and a black hole results from the collapse then it follows from the area theorem that the area of the apparent horizon $\mathcal{A}$ is smaller than the area of the horizon of the final Schwarzschild black hole. Thus an upper bound $E_{\max}$ to the energy emitted E_{rad} as gravitational radiation is

$$E_{\mathrm{rad}} \leq M - \sqrt{\frac{\mathcal{A}}{16\pi}} = E_{\max}. \tag{5.39}$$

As formulated by Thorne the hoop conjecture (5.35) is vague in the sense that the type of horizon is not specified and various interpretations can be given to the mass and to the circumference of the hoop. However we can demonstrate that by using again the collapse of a null shell it is possible to obtain a more precise formulation of this conjecture. Two inequalities of classical differential geometry will play a role here analogous to the Minkowskian inequality in the derivation of (5.34). For any compact and convex domain $\mathcal{D}$ of Euclidean 3-space with boundary $\partial \mathcal{D}$ two separate sets of planar curves are constructed by (i) intersecting $\mathcal{D}$ with any 2-plane and by (ii) considering the boundary of the orthogonal projection of $\mathcal{D}$ onto an arbitrary plane. Let us respectively call L and l the maximum value of the perimeter of these curves. It was shown by Barrabès et al. (1992) that

$$\pi L \leq \int K \, dS \leq 4l, \tag{5.40}$$

where K is the mean curvature at any point of $\partial \mathcal{D}$. Using again the Penrose model of a collapsing shell, and in particular the equation (5.37), one finds that

$$\pi L \leq 16\pi M \leq 4l. \tag{5.41}$$

A similar result was obtained by Tod (1992) and later discussed by Pelath et al. (1998). In contrast to the original formulation of the hoop conjecture both the horizons and the hoop are now given a precise definition. Furthermore the inequality (5.41) provides a justification of the 'when' and 'only when' parts of the hoop conjecture. On the one hand, it states that a necessary condition for the formation of an apparent horizon is that any matter distribution with mass M gets compacted into a region whose plane section in every direction has a maximum perimeter L such that $L \leq 16M$, and on the other hand, that a sufficient condition for the formation of an apparent horizon is that any matter distribution with mass M gets compacted into a region whose orthogonal plane projection has maximum perimeter l such that $l \leq 4\pi M$. This last inequality corresponds to the original formulation of the hoop conjecture (5.35).

We now illustrate these results with a few examples. The simplest example is a spherical shell. In this case $K = 2/R$, $\mu = 1/8\pi R$, $R = 2M$ and since $\mathcal{A} = 4\pi R^2$ and $L = l = 2\pi R$ it immediately follows that $\mathcal{A} = 4\pi(2M)^2$, $L - 16M < 0$, and $l = 2\pi(2M)$, which agrees with (5.34) and (5.41). Furthermore the equation (5.39) shows that $E_{\max}$ vanishes and thus, as expected, no gravitational radiation is emitted during a spherical collapse. Let us now take for $\mathcal{S}$ a cylinder of length d, and with two hemispherical caps of radius R at either end, which collapses radially inwards towards its axis. On the cylinder part we have $K_{\text{cyl}} = 1/R$ and on the hemispherical ends $K_{\text{sph}} = 2/R$. Then by (5.37) one deduces that $\mu_{\text{cyl}} = 1/16\pi R$, $\mu_{\text{sph}} = 1/8\pi R$ and one gets for the total mass of the shell, at the moment of formation of the apparent horizon,

$$M = \frac{R}{2} + \frac{d}{8}. \qquad (5.42)$$

On the other hand

$$\mathcal{A} = 4\pi R^2 + 2\pi R d \quad \text{and} \quad L = l = 2\pi R + 2d. \qquad (5.43)$$

From these values it follows that (5.34) and (5.41) are satisfied since they give

$$\mathcal{A} - 4\pi(2M)^2 < 0, \quad L - 16M < 0, \quad \text{and} \quad l - 2\pi(2M) > 0. \qquad (5.44)$$

Also by (5.39) one sees that gravitational radiation is emitted during the collapse since $E_{\max} > 0$. The condition for the formation of an apparent horizon before the cylinder collapses to a spindle is $R > 0$, which using (5.42) implies that $d < 8M$. Hence if the cylinder is long enough it will collapse to a spindle singularity before an apparent horizon forms. However one expects that the subsequent contraction along the axis ultimately leads to a Schwarzschild black hole.

The Penrose model can be adapted to the collapse of a loop. For example, Hawking (1990) considered a planar circular loop radially collapsing with the speed of light from infinity in Minkowskian space–time. If we assume that the loop lies in the $z = 0$ plane of the rectangular Cartesian coordinates (x, y, z) then the space–time history of the loop is a null 2-space $\mathcal{L}$ lying on the null-cone $\mathcal{N}$ whose internal geometry is everywhere Minkowskian. The equations of $\mathcal{L}$ are $t + r = 0$, $z = 0$ and the 2-space $\mathcal{L}$ is represented in Figure 5.4 by the two lines joining the points C and C' to the vertex of the cone. The segment CC' corresponds to the disk $\mathcal{D}$ in the figure bounded by the loop.

To make use of the Penrose model requires a 2-surface playing the same role as the 2-surface $\mathcal{S}$ above, in particular having a space–time history lying on $\mathcal{N}$, and to

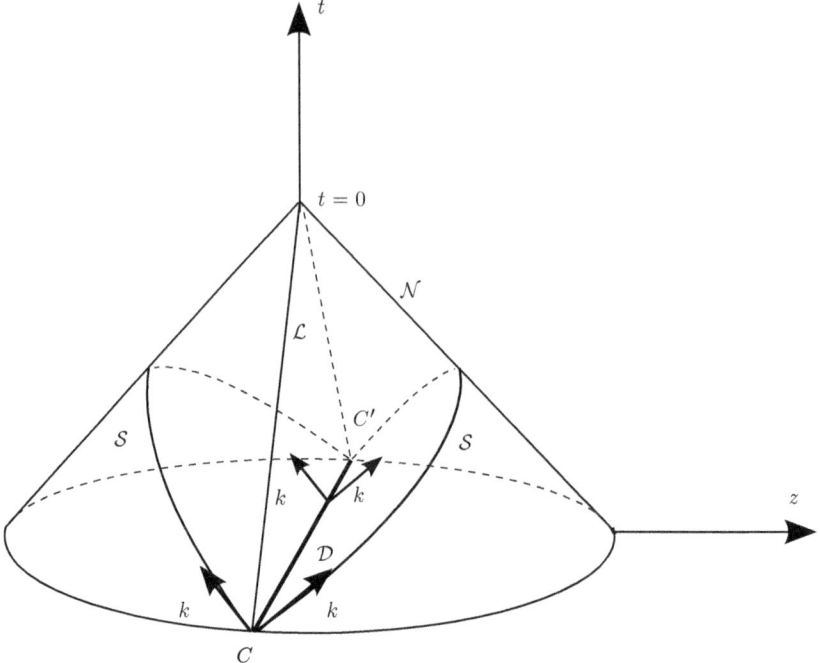

Fig. 5.4 Collapse of a circular loop.

define the outgoing null vectors k orthogonal to $\mathcal{S}$. Let us consider the light rays perpendicular to the disk $\mathcal{D}$. They are expansion-free and span two null planes which intersect the null cone along a 2-surface which we can take as the 2-surface $\mathcal{S}$. The matter content of $\mathcal{S}$ is then concentrated at the cusps C and C'. The inner value of the expansion of the vector field k vanishes everywhere on $\mathcal{S}$ except at the cusps, while its outer value vanishes everywhere on $\mathcal{S}$ when an apparent horizon forms on $\mathcal{S}$, say at time $t = -a$. Let us now derive the inner value of the expansion. The null vector tangent to the generators of $\mathcal{N}$ has components $n^\alpha = (-1, 0, 0, 1)$. The equations of $\mathcal{S}$ are given by $t + r = 0$, $t + a = |z|$, from which one obtains

$$r = \frac{a}{1 + |\cos\theta|} \equiv f(\theta). \tag{5.45}$$

The components of the vector field k derived from the conditions $k \cdot k = k \cdot e_\theta = k \cdot e_\phi = 0$ and $k \cdot n = -1$ are

$$k_a = (1 - F, -\partial_\theta f, 0, -F), \tag{5.46}$$

with $F = (1 - \nabla f . \nabla f)/2$. The expansion of the vector field k is

$$\theta = \frac{1}{f}(1 - {}^2\Delta \ln f), \tag{5.47}$$

where $^2\Delta$ is the 2-dimensional Laplacian calculated on the unit 2-sphere with the spherical polar coordinates (θ, ϕ). Replacing $f(\theta)$ by its expression (5.45) one gets, for the expansion of the outgoing null geodesics,

$$\theta = \frac{1}{a}\delta\left(\theta - \frac{\pi}{2}\right), \qquad (5.48)$$

and by (5.37) the mass M of the 2-surface $\mathcal{S}$ is given by

$$M = \frac{a}{2}. \qquad (5.49)$$

As the area of the apparent horizon is twice the area of the disk $\mathcal{D}$ (i.e. $\mathcal{A} = 2\pi a^2$) we find that

$$\mathcal{A} = 8\pi M^2, \qquad (5.50)$$

which is in agreement with the Gibbons–Penrose isoperimetric inequality. Also since $L = l = 2\pi a$ we obtain $L - 16 < M$ and $l - 2\pi M = 0$ which satisfies the inequalities (5.41). During the collapse of a circular loop gravitational radiation is emitted and the radiated energy has, by (5.39), a maximum value equal to $E_{\max} = M(1 - 1/\sqrt{2})$.

5.3 Scattering properties of high-speed Kerr black holes

A light-like boost of the Kerr gravitational field in an arbitrary direction facilitates the calculation of the angles of deflection of high-speed test particles in the Kerr gravitational field. Relative to the high-speed particles the Kerr gravitational field resembles an impulsive gravitational wave with a singular point on its wavefront (a legacy of the isolated source of the gravitational field). There is a circular disk centred on the singular point which has the property that photons colliding head-on with the gravitational wave and within the disk are reflected backwards and travel with the wave.

To obtain a suitable form of the Kerr solution of Einstein's vacuum field equations for use in this section we begin with the Kerr line-element in the Kerr–Schild form (Kerr 1963, Kerr and Schild 1965a, 1965b) in (5.28) above. However since we will subsequently be applying a Lorentz boost to the solution [see (5.72) below] it will be convenient, for clarity, to write the Kerr line-element out explicitly in barred coordinates as

$$ds^2 = \bar{g}_{ij}\,d\bar{x}^i\,d\bar{x}^j = ds_0^2 + \frac{2M\,\bar{r}^3}{\bar{r}^4 + A^2\,\bar{z}^2}\,(\bar{k}_i\,d\bar{x}^i)^2, \qquad (5.51)$$

with

$$ds_0^2 = \bar{\eta}_{ij}\,d\bar{x}^i\,d\bar{x}^j = d\bar{x}^2 + d\bar{y}^2 + d\bar{z}^2 - d\bar{t}^2, \qquad (5.52)$$

and

$$\bar{k}_i\,d\bar{x}^i = \left(\frac{\bar{r}\,\bar{x} + A\,\bar{y}}{\bar{r}^2 + A^2}\right)d\bar{x} + \left(\frac{\bar{r}\,\bar{y} - A\,\bar{x}}{\bar{r}^2 + A^2}\right)d\bar{y} + \frac{\bar{z}}{\bar{r}}\,d\bar{z} - d\bar{t}. \qquad (5.53)$$

The bars on the coordinates are for convenience and will disappear below. In (5.51) the constant M is the mass of the source and $\mathbf{J} = (0, 0, M\,A)$ is the constant angular momentum of the source. The coordinate $\bar{r}$ is a function of the coordinates $\bar{x}, \bar{y}, \bar{z}$ given by

$$\frac{\bar{x}^2 + \bar{y}^2}{\bar{r}^2 + A^2} + \frac{\bar{z}^2}{\bar{r}^2} = 1. \tag{5.54}$$

The pioneering paper on light-like boosts of gravitational fields is the light-like boost of the Schwarzschild field by Aichelburg and Sexl (1971). We have developed an approach to this subject based on the Riemann curvature tensor (Barrabès and Hogan 2001, 2003a, 2004a) and so we shall require the Riemann tensor for the Kerr space–time in the present context. In the barred coordinates this is given via

$$^{+}\bar{R}_{ijkl} = \bar{R}_{ijkl} + i^{*}\bar{R}_{ijkl}, \tag{5.55}$$

where $^{*}\bar{R}_{ijkl} = \frac{1}{2}\bar{\eta}_{ijpq}\bar{R}^{pq}{}_{kl}$ with $\bar{\eta}_{ijpq} = \sqrt{-\bar{g}}\,\bar{\epsilon}_{ijpq}$ are the components of the left dual of the Riemann tensor (the left and right duals being equal in a vacuum space–time). Specifically in the Kerr case this reads

$$^{+}\bar{R}_{ijkl} = -\frac{M\,\bar{r}^3}{(\bar{r}^2 + i\,A\,\bar{z})^3}\left(\bar{g}_{ijkl} + i\,\bar{\epsilon}_{ijkl} + 3\,\bar{W}_{ij}\,\bar{W}_{kl}\right), \tag{5.56}$$

with

$$\bar{g}_{ijkl} = \bar{g}_{ik}\,\bar{g}_{jl} - \bar{g}_{il}\,\bar{g}_{jk}, \tag{5.57}$$

and $\bar{W}_{ij} = -\bar{W}_{ji}$ given via the 2-form

$$\frac{1}{2}\bar{W}_{ij}\,d\bar{x}^i \wedge d\bar{x}^j = \frac{\bar{r}}{\bar{r}^2 + i\,A\,\bar{z}}\bigg[\bar{x}\,(d\bar{x} \wedge d\bar{t} - i\,d\bar{y} \wedge d\bar{z})$$
$$+\bar{y}\,(d\bar{y} \wedge d\bar{t} - i\,d\bar{z} \wedge d\bar{x})$$
$$+(\bar{z} + i\,A)\,(d\bar{z} \wedge d\bar{t} - i\,d\bar{x} \wedge d\bar{y})\bigg]. \tag{5.58}$$

We shall require the Kerr solution when the angular momentum points in an arbitrary direction in space and not in the positive $\bar{z}$-direction as it does here. Such a form can be found in (3.128) above. However that form is not convenient for our purposes now. Instead, with a, b, c real constants such that $\sqrt{a^2 + b^2 + c^2} = A$, we perform the *rotation*

$$\bar{x} \rightarrow -\frac{(a\,c^2 + b^2 A)}{A\,(b^2 + c^2)}\,\bar{x} - \frac{b\,c\,(a - A)}{A\,(b^2 + c^2)}\,\bar{y} + \frac{c}{A}\,\bar{z}, \tag{5.59}$$

$$\bar{y} \rightarrow -\frac{b\,c\,(a - A)}{A\,(b^2 + c^2)}\,\bar{x} - \frac{(a\,b^2 + c^2 A)}{A\,(b^2 + c^2)}\,\bar{y} + \frac{b}{A}\,\bar{z}, \tag{5.60}$$

$$\bar{z} \rightarrow \frac{c}{A}\,\bar{x} + \frac{b}{A}\,\bar{y} + \frac{a}{A}\,\bar{z}. \tag{5.61}$$

Under this rotation $\mathbf{J} \to (M\,c, M\,b, M\,a)$ and thus the angular momentum of the source points in an arbitrary direction relative to the new $\bar{x}, \bar{y}, \bar{z}$ axes. With (5.59)–(5.61) the function $\bar{r}$ in (5.54) now depends upon the new barred coordinates according to the equation

$$\bar{r}^4 + (a^2 + b^2 + c^2 - \bar{x}^2 - \bar{y}^2 - \bar{z}^2)\,\bar{r}^2 = (c\,\bar{x} + b\,\bar{y} + a\,\bar{z})^2. \tag{5.62}$$

Also the following transformations are consequences of (5.59)–(5.61):

$$d\bar{x}^2 + d\bar{y}^2 + d\bar{z}^2 \to d\bar{x}^2 + d\bar{y}^2 + d\bar{z}^2, \tag{5.63}$$

$$\bar{x}\,d\bar{x} + \bar{y}\,d\bar{y} + (\bar{z} + i\,A)\,d\bar{z} \to (\bar{x} + i\,c)\,d\bar{x} + (\bar{y} + i\,b)\,d\bar{y}$$
$$+ (\bar{z} + i\,a)\,d\bar{z}, \tag{5.64}$$

$$\bar{y}\,d\bar{x} - \bar{x}\,d\bar{y} \to \frac{(a\,\bar{y} - b\,\bar{z})}{A}\,d\bar{x} + \frac{(c\,\bar{z} - a\,\bar{x})}{A}\,d\bar{y} + \frac{(b\,\bar{x} - c\,\bar{y})}{A}\,d\bar{z}, \tag{5.65}$$

and

$$\bar{x}\,d\bar{y} \wedge d\bar{z} + \bar{y}\,d\bar{z} \wedge d\bar{x} + (\bar{z} + i\,A)\,d\bar{x} \wedge d\bar{y}$$
$$\to (\bar{x} + i\,c)\,d\bar{y} \wedge d\bar{z} + (\bar{y} + i\,b)\,d\bar{z} \wedge d\bar{x} + (\bar{z} + i\,a)\,d\bar{x} \wedge d\bar{y}. \tag{5.66}$$

It thus follows that under the rotation (5.59)–(5.61) the Kerr metric tensor retains the Kerr–Schild form so that

$$\bar{g}_{ij} = \bar{\eta}_{ij} + 2\,H\,\bar{k}_i\,\bar{k}_j, \tag{5.67}$$

with $\bar{\eta}_{ij} = \mathrm{diag}(1,1,1,-1)$,

$$H = \frac{M\,\bar{r}^3}{\bar{r}^4 + (\mathbf{a}\cdot\bar{\mathbf{x}})^2}, \tag{5.68}$$

and

$$\bar{k}_i\,d\bar{x}^i = \frac{(\mathbf{a}\cdot\bar{\mathbf{x}})(\mathbf{a}\cdot d\bar{\mathbf{x}})}{\bar{r}\,(\bar{r}^2 + |\mathbf{a}|^2)} + \frac{\bar{r}\,(\bar{\mathbf{x}}\cdot d\bar{\mathbf{x}})}{\bar{r}^2 + |\mathbf{a}|^2} - \frac{\mathbf{a}\cdot(\bar{\mathbf{x}} \times d\bar{\mathbf{x}})}{\bar{r}^2 + |\mathbf{a}|^2} - d\bar{t}. \tag{5.69}$$

In these formulae $\mathbf{a} = (c,b,a)$ with $|\mathbf{a}|^2 = a^2 + b^2 + c^2$ and $\bar{\mathbf{x}} = (\bar{x},\bar{y},\bar{z})$ with $\bar{r}$ given in terms of $\bar{x},\bar{y},\bar{z}$ by (5.62). The centre dot and cross denote the usual scalar and vector products in 3-dimensional Euclidean space. This form of the Kerr line-element has been given by Weinberg (1972) with Weinberg's angular momentum pointing in the opposite direction to ours. The Riemann curvature tensor is now given by

$$^+\bar{R}_{ijkl} = -\frac{M\,\bar{r}^3}{(\bar{r}^2 + i\,(\mathbf{a}\cdot\bar{\mathbf{x}}))^3}\left(\bar{g}_{ijkl} + i\,\bar{\epsilon}_{ijkl} + 3\,\bar{W}_{ij}\,\bar{W}_{kl}\right), \tag{5.70}$$

with $\bar{g}_{ijkl}$ given now by (5.57) and (5.67) and $\bar{W}_{ij}$ by

$$\frac{1}{2}\bar{W}_{ij}\,d\bar{x}^i \wedge d\bar{x}^j = \frac{\bar{r}}{\bar{r}^2 + i\,(\mathbf{a}\cdot\bar{\mathbf{x}})}\Big[(\bar{x} + i\,c)\,(d\bar{x} \wedge d\bar{t} - i\,d\bar{y} \wedge d\bar{z})$$
$$+ (\bar{y} + i\,b)\,(d\bar{y} \wedge d\bar{t} - i\,d\bar{z} \wedge d\bar{x})$$
$$+ (\bar{z} + i\,a)\,(d\bar{z} \wedge d\bar{t} - i\,d\bar{x} \wedge d\bar{y})\Big]. \tag{5.71}$$

94 Black holes

We now make a Lorentz boost in the $\bar{x}$-direction given by (remembering that we use units for which $c = 1$)

$$\bar{x} = \gamma(x - v t), \quad \bar{y} = y, \quad \bar{z} = z, \quad \bar{t} = \gamma(t - v x), \tag{5.72}$$

with $\gamma = (1 - v^2)^{-1/2}$. In carrying out this boost we make the assumption [as in for example Aichelburg and Sexl (1971)] that the relative energy p of the Kerr source remains fixed and thus the rest-mass scales as $M = p\gamma^{-1}$. In addition we assume that the components of the angular momentum per unit mass orthogonal to the boost direction (the constants b and a) remain constant but the component c in the boost direction scales as $c = \hat{c}\gamma^{-1}$, where $\hat{c}$ is a constant [see Barrabès and Hogan (2003a) for an explanation of this]. The physical explanation of the scaling of c is that the multipole moments of the isolated source of a gravitational field suffer a Lorentz contraction in the boost direction (Barrabès and Hogan 2001). In the Kerr case the multipole moments are constructed from the mass and the angular momentum per unit mass in a manner described by Zel'dovich and Novikov (1978) in such a way that the Lorentz contraction in the direction of the boost is equivalent to the scaling of c in that direction given here. When (5.72) is applied to the Riemann tensor (5.70) we obtain the Riemann tensor components denoted $^+R_{ijkl}$ without a bar. We then take the light-like limit $v \to 1$ to obtain the Riemann tensor components denoted $^+\tilde{R}_{ijkl}$. The latter are therefore given by

$$^+\tilde{R}_{ijkl} = \lim_{v \to 1} {}^+R_{ijkl}. \tag{5.73}$$

To evaluate this limit we need the following: first we have the identity

$$\frac{\bar{r}^3}{[\bar{r}^2 + i(\mathbf{a} \cdot \bar{\mathbf{x}})]^3} = \frac{1}{[(\bar{y} + i b)^2 + (\bar{z} + i a)^2]} \frac{\partial}{\partial \bar{x}} \left(\frac{(\bar{x} + i c)\bar{r}}{\bar{r}^2 + i(\mathbf{a} \cdot \bar{\mathbf{x}})} \right). \tag{5.74}$$

We derive from this [the reader may wish to consult Barrabès and Hogan (2003a) for assistance] the limit

$$\lim_{v \to 1} \frac{\gamma \bar{r}^3}{[\bar{r}^2 + i(\mathbf{a} \cdot \bar{\mathbf{x}})]^3} = \frac{2\delta(x + t)}{(y + i b)^2 + (z + i a)^2}, \tag{5.75}$$

where $\delta(x + t)$ is the Dirac delta function which is singular on $x + t = 0$. Differentiating (5.75) with respect to $\bar{y} = y$ and using

$$\frac{\partial \bar{r}}{\partial \bar{y}} = \frac{\bar{r}^3 \bar{y} + \bar{r} b (\mathbf{a} \cdot \bar{\mathbf{x}})}{\bar{r}^4 + i(\mathbf{a} \cdot \bar{\mathbf{x}})^2}, \tag{5.76}$$

we obtain another useful limit:

$$\lim_{v \to 1} \frac{\gamma \bar{r}^5}{[\bar{r}^2 + i(\mathbf{a} \cdot \bar{\mathbf{x}})]^5} = \frac{4}{3} \frac{\delta(x + t)}{[(y + i b)^2 + (z + i a)^2]^2}. \tag{5.77}$$

Now evaluating (5.73) we find, for example,

$$^+\tilde{R}_{1212} = 4\,p\left\{\frac{z+i\,a+i\,(y+i\,b)}{(z+i\,a)^2+(y+i\,b)^2}\right\}^2 \delta(x+t),$$

$$= 4\,p\left\{\frac{z+b+i\,(y-a)}{(z+b)^2+(y-a)^2}\right\}^2 \delta(x+t). \tag{5.78}$$

We can rewrite this in the form

$$^+\tilde{R}_{1212} = (h_{yy} - i h_{yz})\,\delta(x+t), \tag{5.79}$$

with

$$h = 2\,p\,\log\{(y-a)^2 + (z+b)^2\}, \tag{5.80}$$

and the subscripts on h in (5.79) denote partial derivatives. When all of the components of the light-like boosted Riemann tensor (5.73) are calculated in this way we find that $\tilde{R}_{ijkl} \equiv 0$ except for

$$^+\tilde{R}_{1212} = {}^+\tilde{R}_{2424} = -{}^+\tilde{R}_{1313} = -{}^+\tilde{R}_{3434} = -{}^+\tilde{R}_{3134} = {}^+\tilde{R}_{2124}$$
$$= (h_{yy} - i\,h_{yz})\,\delta(x+t), \tag{5.81}$$

$$^+\tilde{R}_{1213} = {}^+\tilde{R}_{2434} = {}^+\tilde{R}_{3124} = {}^+\tilde{R}_{2134}$$
$$= i\,(h_{yy} - i\,h_{yz})\,\delta(x+t), \tag{5.82}$$

with h given by (5.80). This Riemann curvature tensor can be obtained from the metric tensor given via the line-element

$$ds^2 = dx^2 + dy^2 + dz^2 - dt^2 = 2h\,\delta(x+t)\,(dx+dt)^2. \tag{5.83}$$

We have arrived here at a space–time model of the gravitational field of a plane, inhomogeneous, impulsive gravitational wave with the null hyperplane $x+t=0$ as the history of the wavefront. The curvature tensor has a delta function singularity (reflecting the profile of the wave) and is also singular on the null geodesic generator $y=a, z=-b$ of the null hyperplane $x+y=0$. The Aichelburg–Sexl (1971) result for the boosted Schwarzschild solution is obtained by putting the angular momentum parameters b and a to zero. Since the angular momentum per unit mass c in the boost direction scales differently from the transverse components b and a in terms of the 3-velocity v of the observer in (5.72) it does not appear in the light-like limit $v \to 1$. Hence the effect of the presence of the angular momentum is simply to shift the singularity on the wavefront (a legacy of the original isolated source) from $y=z=0$ in the Schwarzschild case to $y=a, z=-b$ for the Kerr field with angular momentum $\mathbf{J} = (M\,c, M\,b, M\,a)$ if the light-like boost is in the x-direction.

The reader may wish to compare the approach to the light-like boost of the Kerr field here with the point of view described in Balasin and Nachbagauer (1996). There the angle α between the axis of symmetry of the Kerr field and the direction of boost is introduced and the authors conclude that from their point of view 'only the limiting cases $\alpha = 0$ and $\alpha = \pi/2$......admit a solution in closed form'. In addition they point out that in their analysis 'the general case allows a perturbative treatment if one suitably rescales the coordinates and expands the resulting expression with respect to $\sin \alpha$'. We have avoided these restrictions by first tilting the axis of the Kerr source and then boosting in the $\bar{x}$-direction. The positive advantages of our *Riemann tensor centered approach* are emphasized in Barrabès and Hogan (2003a, 2003b).

Let $\bar{S}$ be the rest-frame of the Kerr source and S the rest-frame of a high-speed particle projected into the Kerr field. From the point of view of S the Kerr gravitational field resembles that of an impulsive gravitational wave modelled by the space–time with line-element (5.83). In S the world line of the particle is a time-like geodesic of (5.83). In S we assume that the particle is located at $x = 0, y = y_0, z = z_0$. It starts moving after encountering the gravitational wave. Using the time-like geodesic equations of the space–time with line-element (5.83) we can obtain the 4-velocity of the particle before and after encountering the impulsive gravitational wave, calculated in the frame S (Barrabès and Hogan 2004b). The components of the 4-velocity of the particle before and after scattering in the frame $\bar{S}$ are obtained by Lorentz transformation (5.72) with v close to unity. If α is the deflection angle out of the $\bar{x}\bar{z}$-plane of the high-speed particle after encountering the gravitational wave and if β is the deflection angle out of the $\bar{x}\bar{y}$-plane then these angles are given by

$$\tan \alpha = \frac{\dot{y}}{\sqrt{\dot{x}^2 + \dot{y}^2}} = \frac{Y_1}{\{Z_1^2 + \gamma^2 [X_1 (1-v) + v]^2\}^{1/2}}, \tag{5.84}$$

$$\tan \beta = \frac{\dot{z}}{\dot{x}} = \frac{Z_1}{\gamma \{X_1 (1-v) + v\}}, \tag{5.85}$$

with the dots indicating differentiation with respect to proper time and X_1, Y_1, Z_1 calculated from (5.80) according to

$$X_1 = -\frac{1}{2} \left[(h_y)_0^2 + (h_z)_0^2\right], \tag{5.86}$$

$$Y_1 = -(h_y)_0, \tag{5.87}$$

$$Z_1 = -(h_z)_0. \tag{5.88}$$

The brackets around the partial derivatives followed by a subscript zero here denote that the quantity is calculated at $y = y_0, z = z_0$. Thus for v close to unity we find that

$$\tan \alpha = \frac{-4 M (y_0 - a)}{[\{(y_0 - a)^2 + (z_0 + b)^2 - 4 M^2\}^2 + 16 M^2 (z_0 + b)^2]^{1/2}}, \tag{5.89}$$

and

$$\tan \beta = \frac{-4 M (z_0 + b)}{(y_0 - a)^2 + (z_0 + b)^2 - 4 M^2}. \tag{5.90}$$

If the projected particle starts at $y_0 = -\eta$ (with $\eta > 0$), $z_0 = 0$ then for large impact parameter η, (5.89) and (5.90) give the small angles of deflection

$$\alpha = \frac{4M}{\eta} - \frac{4M\,a}{\eta^2}, \qquad (5.91)$$

$$\beta = -\frac{4M\,b}{\eta}. \qquad (5.92)$$

The first of these agrees with the small angle of deflection of a photon moving in the equatorial plane of the Kerr source calculated by Boyer and Lindquist (1967).

It is interesting from a physical and a geometrical point of view to study time-like and null geodesics in the space–time with line-element (5.83). In this way we get a picture of the scattering properties of high-speed particles and photons in the Kerr gravitational field. In particular we shall exhibit the focusing behaviour of a time-like congruence in the space–time with line-element (5.83) and see that it is precisely what one would expect in the field of a rotating source. We then consider the head-on collision of photons with the gravitational wave and demonstrate that some photons are reflected backwards on collision with the wave leading to the phenomenon of the *dimming of the signal*. In addition it can be shown (Barrabès et al. 2005) that if the photons are replaced by electromagnetic waves then it is the high-frequency waves that are reflected by the gravitational wave.

Time-like geodesics in the space–time described by the line-element (5.83) have been studied by Barrabès and Hogan (2004b). Let τ be proper-time along such a geodesic with $\tau < 0$ on the geodesic before it intersects the null hyperplane $x + t = 0$, $\tau = 0$ at the intersection of the geodesic with $x + t = 0$ and $\tau > 0$ after the intersection and to the future of the null hyperplane. For the simplest initial conditions we find that a time-like geodesic is given for $\tau < 0$ by

$$x = x_0 + x_1\,\tau, \qquad (5.93)$$
$$y = y_0, \qquad (5.94)$$
$$z = z_0, \qquad (5.95)$$
$$t = -x_0 + t_1\,\tau, \qquad (5.96)$$

and for $\tau > 0$ by

$$x = x_0 + x_1\,\tau + X_1\,\tau + \hat{X}_1, \qquad (5.97)$$
$$y = y_0 + Y_1\,\tau, \qquad (5.98)$$
$$z = z_0 + Z_1\,\tau, \qquad (5.99)$$
$$t = -x_0 + t_1\,\tau - X_1\,\tau - \hat{X}_1, \qquad (5.100)$$

with $\hat{X}_1 = (h)_0$ and X_1, Y_1, Z_1 given by (5.86)–(5.88). Also $x_1 + t_1 = 1$ and $x_1 - t_1 = -1$. We have here two time-like congruences parametrized by x_0, y_0, z_0, one for $\tau < 0$ and the more interesting one for $\tau > 0$. The unit time-like tangent to the congruence on the future side of $\tau = 0$ has components in coordinates x, y, z, t given by

$$v^i = (x_1 + X_1,\, Y_1,\, Z_1,\, t_1 - X_1). \qquad (5.101)$$

We can consider x_0, y_0, z_0, τ as scalar fields defined by (5.97)–(5.100) on the region of Minkowskian space–time $\tau \geq 0$. Hence v^i becomes a vector field in the region $\tau \geq 0$ whose integral curves constitute the time-like congruence we wish to study.

98 Black holes

In coordinates $x^i = (\zeta, \bar\zeta, x_0, \tau)$ with $\zeta = y_0 + iz_0$ we have

$$v_i\, dx^i = dx_0 - d\tau \quad \text{and} \quad v^i \frac{\partial}{\partial x^i} = \frac{\partial}{\partial \tau}, \tag{5.102}$$

and the line-element of Minkowskian space–time for $\tau \geq 0$ reads

$$ds^2 = \left| d\zeta - 2\tau\, \frac{\partial^2 (h)_0}{\partial \bar\zeta^2}\, d\bar\zeta \right|^2 + 2\, d\tau\, dx_0 - d\tau^2. \tag{5.103}$$

In the region of Minkowskian space–time $\tau < 0$ the time-like congruence is given by (5.93)–(5.96). In the coordinates $x^i = (\zeta, \bar\zeta, x_0, \tau)$ the line-element can be written for $\tau < 0$ as

$$ds^2 = |d\zeta|^2 + 2\, d\tau\, dx_0 - d\tau^2. \tag{5.104}$$

Using the Heaviside step function $\vartheta(\tau)$, which as before is unity if $\tau > 0$ and vanishes if $\tau < 0$, we can combine the line-elements (5.103) and (5.104) into one convenient formula:

$$ds^2 = \left| d\zeta - 2\tau\, \vartheta(\tau)\, \frac{\partial^2 (h)_0}{\partial \bar\zeta^2}\, d\bar\zeta \right|^2 + 2\, d\tau\, dx_0 - d\tau^2. \tag{5.105}$$

Calculation of the Riemann and Ricci tensors directly from this reveals that the Riemann tensor has a delta function singularity on $\tau = 0$ (since $\tau = 0$ is the history of an impulsive gravitational wave) and the Ricci tensor vanishes for all values of τ, in particular for $\tau = 0$.

To calculate the properties of the time-like congruence of integral curves of the vector field v^i given by (5.102) for $\tau > 0$ we require the components of the covariant derivative of v^i (or v_i). This is indicated by a semicolon and is found to be expressible as

$$v_{i;j} = \hat\mu_1\, n_{(1)i}\, n_{(1)j} + \hat\mu_2\, n_{(2)i}\, n_{(2)j} = v_{j;i}, \tag{5.106}$$

with $n_{(1)i}, n_{(2)i}$ given via the 1-forms

$$n_{(1)i}\, dx^i = \frac{i}{2}(1 + 2\tau |h_{\zeta\zeta}|) \left\{ \left(\frac{h_{\zeta\zeta}}{|h_{\zeta\zeta}|}\right)^{1/2} d\zeta - \left(\frac{|h_{\zeta\zeta}|}{h_{\zeta\zeta}}\right)^{1/2} d\bar\zeta \right\}, \tag{5.107}$$

$$n_{(2)i}\, dx^i = \frac{i}{2}(1 - 2\tau |h_{\zeta\zeta}|) \left\{ \left(\frac{h_{\zeta\zeta}}{|h_{\zeta\zeta}|}\right)^{1/2} d\zeta + \left(\frac{|h_{\zeta\zeta}|}{h_{\zeta\zeta}}\right)^{1/2} d\bar\zeta \right\}. \tag{5.108}$$

We emphasize that $h_{\zeta\zeta} = \partial^2 (h)_0 / \partial \zeta^2$ here. The vectors $n_{(1)i}, n_{(2)i}$ are unit space-like vectors orthogonal to each other and to v^i and are parallel transported along the integral curves of v^i. Clearly from (5.106) the geodesic congruence is twist-free. The scalars $\hat\mu_1, \hat\mu_2$ are given by

$$\hat\mu_1 = \frac{2|h_{\zeta\zeta}|}{1 + 2\tau |h_{\zeta\zeta}|}, \tag{5.109}$$

$$\hat{\mu}_2 = \frac{-2|h_{\zeta\zeta}|}{1 - 2\tau|h_{\zeta\zeta}|}, \qquad (5.110)$$

and thus the contraction of the congruence is

$$\theta = v^i{}_{;i} = \hat{\mu}_1 + \hat{\mu}_2 = -\frac{8\tau|h_{\zeta\zeta}|^2}{1 - 4\tau^2|h_{\zeta\zeta}|^2}. \qquad (5.111)$$

The shear tensor associated with the congruence is now

$$\sigma_{ij} = v_{i;j} - \frac{1}{3}\theta\,(g_{ij} + v_i\,v_j). \qquad (5.112)$$

A useful orthonormal tetrad which is parallel transported along the congruence is $\{n^i_{(1)}, n^i_{(2)}, n^i, v^i\}$ with n^i given by

$$n^i \frac{\partial}{\partial x^i} = \frac{\partial}{\partial x_0} + \frac{\partial}{\partial \tau}. \qquad (5.113)$$

We can write the shear tensor in terms of this orthonormal tetrad as

$$\sigma_{ij} = \mu_1\,n_{(1)i}\,n_{(1)j} + \mu_2\,n_{(2)i}\,n_{(2)j} + \mu_3\,n_i\,n_j, \qquad (5.114)$$

with

$$\mu_1 = \hat{\mu}_1 - \frac{1}{3}\theta, \quad \mu_2 = \hat{\mu}_2 - \frac{1}{3}\theta, \quad \mu_3 = -\frac{1}{3}\theta. \qquad (5.115)$$

The orthonormal tetrad vectors $\{n^i_{(1)}, n^i_{(2)}, n^i, v^i\}$ are eigenvectors of the shear tensor with corresponding eigenvalues $\mu_1, \mu_2, \mu_3, 0$. From (5.111) and (5.115) we see that $\mu_1 + \mu_2 + \mu_3 = 0$, confirming that $g^{ij}\sigma_{ij} = 0$ as it should be. The time-like congruence approximates the world lines of high-speed particles after they have been deflected by the rotating black hole. By (5.111) the lines of the congruence converge at $\tau = \tau_0$ given by

$$\tau_0 = \frac{(y_0 - a)^2 + (z_0 + b)^2}{4\,p}. \qquad (5.116)$$

But (5.98) and (5.99) give

$$y - a = (y_0 - a)\left(1 - \frac{\tau}{\tau_0}\right), \qquad (5.117)$$

$$z + b = (z_0 + b)\left(1 - \frac{\tau}{\tau_0}\right), \qquad (5.118)$$

demonstrating that the right-hand sides vanish at $\tau = \tau_0$. Hence *the paths of the high-speed particles converge on* a straight line after being scattered by the black hole and this straight line is *the intersection of the planes $y = a, z = -b$.*

The head-on collision of photons with the gravitational wave having $\tau = 0$ as its history is an interesting study. In the region of Minkowskian space–time $\tau < 0$, to the past of the history of the impulsive gravitational wave, we consider a congruence of null geodesics tangent to the null vector field (in coordinates x, y, z, t) with components

$$^{(-)}l^i = (1, 0, 0, 1). \qquad (5.119)$$

100 Black holes

The superscript minus on a quantity will denote its value prior to the collision with the gravitational wave. We shall find useful a null tetrad composed of $^{(-)}l^i, ^{(-)}n^i, ^{(-)}m^i$, and $^{(-)}\bar{m}^i$ with

$$^{(-)}n^i = \frac{1}{2}(1,0,0,-1) \quad \text{and} \quad ^{(-)}m^i = \frac{1}{\sqrt{2}}(0,1,i,0), \tag{5.120}$$

and $^{(-)}\bar{m}^i$ the complex conjugate of $^{(-)}m^i$. All of the scalar products among the null tetrad vectors defined here vanish with the exception of $^{(-)}n^i\,^{(-)}l_i = 1 = {}^{(-)}\bar{m}^i\,^{(-)}m_i$. Starting with a null geodesic with tangent (5.119) in the region to the past of the null hyperplane $x + t = 0$ we wish to follow it through the null hyperplane into the region to the future of the hyperplane. To do this we must solve the null geodesic equations in the space–time with line-element (5.83). If the parametric equations of the null geodesic are $x^i = x^i(\lambda)$ with λ an affine parameter and $\lambda < 0$ on the null geodesic before encountering the null hyperplane $x + t = 0 \Leftrightarrow \lambda = 0$ and $\lambda > 0$ on the null geodesic after encountering the null hyperplane $x + t = 0$ we find [see Barrabès et al. (2005) for details] that for $\lambda < 0$ the null geodesic is given by

$$x = x_0 + \lambda, \tag{5.121}$$
$$y = y_0, \tag{5.122}$$
$$z = z_0, \tag{5.123}$$
$$t = -x_0 + \lambda, \tag{5.124}$$

with x_0, y_0, z_0 constants of integration, and for $\lambda > 0$ the null geodesic is given by

$$x = x_0 + \lambda + X_1 \lambda + \hat{X}_1, \tag{5.125}$$
$$y = y_0 + Y_1 \lambda, \tag{5.126}$$
$$z = z_0 + Z_1 \lambda, \tag{5.127}$$
$$t = -x_0 + \lambda - X_1 \lambda - \hat{X}_1, \tag{5.128}$$

where now

$$\hat{X}_1 = (h)_0, \quad X_1 = -(h_y)_0^2 - (h_z)_0^2, \quad Y_1 = -2(h_y)_0, \quad Z_1 = -2(h_z)_0, \tag{5.129}$$

with the functions inside the round brackets with the subscript zero evaluated at $y = y_0, z = z_0$. It is clear from an inspection of (5.121)–(5.128) that the point at which the null geodesic in $\lambda < 0$ meets the null hyperplane $\lambda = 0$ and the point from which it leaves the null hyperplane and enters the region $\lambda > 0$ are two different points (both specified by the constants x_0, y_0, z_0) on account of a translation along the generators of $\lambda = 0$ in going from the past side of $\lambda = 0$ to the future side. The tangent to the null geodesic on the future side of $\lambda = 0$ is

$$l^i = (1 + X_1, Y_1, Z_1, 1 - X_1). \tag{5.130}$$

A useful null tetrad defined on the future side of $\lambda = 0$ at the point specified by x_0, y_0, z_0 is given by $l^i, n^i, m^i, \bar{m}^i$ with

$$n^i = \frac{1}{2}(1,0,0,-1) = {}^{(-)}n^i, \qquad (5.131)$$

$$m^i = \frac{1}{\sqrt{2}}\left(-\frac{1}{2}(Y_1 + iZ_1), 1, i, \frac{1}{2}(Y_1 + iZ_1)\right), \qquad (5.132)$$

with $\bar{m}^i$ the complex conjugate of m^i.

To aid the development of the geometrical analysis, the vectors $l^i, n^i, m^i, \bar{m}^i$ defined on the future side of $\lambda = 0$ by (5.130)–(5.132) can be extended to vector fields in the region $\lambda > 0$ by parallel transport along the null geodesics tangent to l^i emanating into the region $\lambda > 0$ from each point on the future side of $\lambda = 0$ specified by x_0, y_0, z_0. We require the derivatives of these vector fields with respect to $x^i = (x, y, z, t)$. To achieve this we note that x_0, y_0, z_0, λ can be extended to scalar fields on the region $\lambda > 0$ by reading (5.125)–(5.128) as defining x_0, y_0, z_0, λ as functions of x, y, z, t. Calculating the derivatives of x_0, y_0, z_0, λ in this way we first find that

$$\frac{\partial \lambda}{\partial x^i} = n_i \quad \text{and} \quad \frac{\partial x_0}{\partial x^i} = \frac{1}{2} l_i. \qquad (5.133)$$

It thus follows that the integral curves of the vector fields n^i, l^i are twist-free null geodesics. The integral curves of the vector field n^i generate the null hyperplanes $\lambda = \text{constant} > 0$ while the integral curves of the vector field l^i generate the null hyper*surfaces* $x_0 = \text{constant}$. The latter is the case since the integral curves of l^i are found below to have contraction and shear. The derivatives of y_0 and z_0 are required and are given by the following: If

$$\varphi = 1 - 4\lambda^2\{(h_{yz})_0^2 + (h_{yy})_0^2\}, \qquad (5.134)$$

then we find that

$$\varphi \frac{\partial y_0}{\partial x} = -\frac{1}{2}\left\{(y_1 + Y_1)\left(1 + \lambda\frac{\partial Z_1}{\partial z_0}\right) - (z_1 + Z_1)\lambda\frac{\partial Y_1}{\partial z_0}\right\} = \varphi\frac{\partial y_0}{\partial t}, \qquad (5.135)$$

$$\varphi \frac{\partial z_0}{\partial x} = -\frac{1}{2}\left\{(z_1 + Z_1)\left(1 + \lambda\frac{\partial Y_1}{\partial y_0}\right) - (y_1 + Y_1)\lambda\frac{\partial Z_1}{\partial y_0}\right\} = \varphi\frac{\partial z_0}{\partial t}, \qquad (5.136)$$

$$\varphi \frac{\partial y_0}{\partial y} = 1 + \lambda\frac{\partial Z_1}{\partial z_0}, \qquad (5.137)$$

$$\varphi \frac{\partial z_0}{\partial y} = -\lambda\frac{\partial Z_1}{\partial y_0}, \qquad (5.138)$$

$$\varphi \frac{\partial y_0}{\partial z} = -\lambda\frac{\partial Y_1}{\partial z_0}, \qquad (5.139)$$

$$\varphi \frac{\partial z_0}{\partial z} = 1 + \lambda\frac{\partial Y_1}{\partial y_0}. \qquad (5.140)$$

Using these formulae we arrive at

$$\sigma = l_{i,j}\, m^i\, m^j = \frac{-4\,\{(h_{yy})_0 + i(h_{yz})_0\}}{1 - 4\lambda^2\,\{(h_{yy})_0^2 + (h_{yz})_0^2\}}, \tag{5.141}$$

and

$$\rho = l_{i,j}\, m^i\, \bar m^j = \frac{-4\lambda\,\{(h_{yy})_0^2 + (h_{yz})_0^2\}}{1 - 4\lambda^2\,\{(h_{yy})_0^2 + (h_{yz})_0^2\}}. \tag{5.142}$$

The scalar σ is the complex shear of the null geodesic congruence tangent to l^i for $\lambda > 0$ and the real scalar ρ is the contraction of this congruence. The vector $^{(-)}l^i$ in (5.119) is a constant vector field in $\lambda < 0$ and thus its integral curves are twist-free, expansion-free, and shear-free null geodesics. We see from (5.141) and (5.142) that on crossing $\lambda = 0$ this congruence experiences a jump in the shear while the contraction is continuous at $\lambda = 0$ (since ρ vanishes on $\lambda = 0$). This characteristic behaviour of a null geodesic congruence intersecting the history of an impulsive gravitational wave was first pointed out by Penrose (1972). Now that we have available in (5.133)–(5.140) the derivatives of x_0, y_0, z_0, λ with respect to x, y, z, t we can evaluate the derivatives of X_1, Y_1, Z_1 in (5.129) and then the derivatives of l^i, n^i, m^i in (5.130)–(5.132). The end result is the neat formulae:

$$l_{i,j} = \bar\sigma\, m_i\, m_j + \sigma\, \bar m_i\, \bar m_j + \rho\,(m_i\, \bar m_j + \bar m_i\, m_j), \tag{5.143}$$
$$n_{i,j} = 0, \tag{5.144}$$
$$m_{i,j} = -\rho\, n_i\, m_j - \sigma\, n_i\, \bar m_j, \tag{5.145}$$

with σ and ρ given by (5.141) and (5.142). From (5.142) it follows that neighbouring null geodesics tangent to l^i intersect when $\lambda = \lambda_0 > 0$ given by $4\lambda_0^2\{(h_{yy})_0^2 + (h_{yz})_0^2\} = 1$. With $h(y,z)$ given by (5.80) the intersection occurs when

$$8\,p\,\lambda_0 = (y_0 - a)^2 + (z_0 + b)^2. \tag{5.146}$$

It follows from (5.126) and (5.127) that

$$y - a = (y_0 - a)\left(1 - \frac{\lambda}{\lambda_0}\right), \tag{5.147}$$

$$z + b = (z_0 + b)\left(1 - \frac{\lambda}{\lambda_0}\right), \tag{5.148}$$

and thus when $\lambda = \lambda_0$ given by (5.146) the integral curves of the vector field l^i focus on the straight line $y = a, z = -b$. That the rays converge on a straight line is due to the fact that the lensing source is moving on a straight line. For the regions of space–time corresponding to $\lambda > 0$ and $\lambda < 0$ we may regard $y^i = (x_0, y_0, z_0, \lambda)$ as 'optical coordinates', in the terminology of Synge (1964), based on the null hyperplane $\lambda = 0$. The relationship between these optical coordinates and the rectangular Cartesian coordinates and time is given by (5.125)–(5.128) for $\lambda > 0$ and by (5.121)–(5.124) for $\lambda < 0$. With $\zeta = y_0 + i z_0$ the line-element of the space–time including the history of the impulsive gravitational wave analogous to (5.105), in coordinates y^i, is found to be

$$ds^2 = \left| d\zeta + 2\lambda\vartheta(\lambda)\frac{\partial Y_1}{\partial\bar\zeta}d\bar\zeta \right|^2 + 4\,d\lambda\,dx_0. \tag{5.149}$$

The 4-momentum of photons prior to head-on collision with the gravitational wave is given by (5.119) and immediately after collision on the future side of the null hyperplane $\lambda = 0$ the 4-momentum is given by (5.130) which we can write explicitly as

$$l^i = \left(1 - \frac{16\,p^2}{R_0^2},\; -\frac{8\,p\,(y_0-a)}{R_0^2},\; -\frac{8\,p\,(z_0+b)}{R_0^2},\; 1 + \frac{16\,p^2}{R_0^2} \right), \tag{5.150}$$

where $R_0^2 = (y_0 - a)^2 + (z_0 + b)^2$. For $R_0 < 4\,p$ this 4-momentum is given approximately by

$$l^i = -\frac{16\,p^2}{R_0^2}(1,0,0,-1) = -\frac{32\,p^2}{R_0^2}\,n^i, \tag{5.151}$$

with n^i given by (5.131) and we note that n^i is past-pointing. It follows from (5.151) that photons colliding with the gravitational wave sufficiently close to the singular point $y_0 = a$, $z_0 = -b$ are reflected back and accompany the gravitational wave which is moving in the reflected direction. Hence there will be observers who will see a circular disk on their sky corresponding to

$$(y_0 - a)^2 + (z_0 + b)^2 \leq 16\,p^2. \tag{5.152}$$

Since this is an approximate result the disk is not opaque to these observers but the light passing through it is *dimmed* since most of it is reflected backwards. When the incoming photons are replaced by monochromatic, plane electromagnetic waves (whose electromagnetic field is considered a test field on the Minkowskian space–time $\lambda < 0$) it is found that only waves of frequency $\omega > p^{-1}$ are reflected and thus fail to pass through the gravitational wave (Barrabès *et al.* 2005).

5.4 Inside the black hole

One of the most famous theorems concerning black holes is the *no-hair theorem* which states that for a generic gravitational collapse, and provided cosmic censorship is valid, the field exterior to the event horizon has the Kerr–Newman form characterized by just three parameters (mass M, charge Q, and angular momentum $J = Ma$). During the collapse perturbations developing at the surface of the star produce emission of gravitational radiation which dies out with advanced time as an inverse-power law (Price 1972a, 1972b) and the end product of the collapse is, in the most general case, a Kerr–Newman black hole. We are interested in this section in the interior of the black hole. Inside the event horizon of a Schwarzschild black hole (i.e. for $r < 2M$) the radial coordinate r becomes time-like and the direction of time coincides with the direction of decreasing r. The singularity at $r = 0$ is a space-like hypersurface and all hypersurfaces $r = \text{constant} < 2M$ are space-like and have topology $S^2 \times R$ (see Figure 5.5). In the case of a charged and/or rotating black hole the singularity at

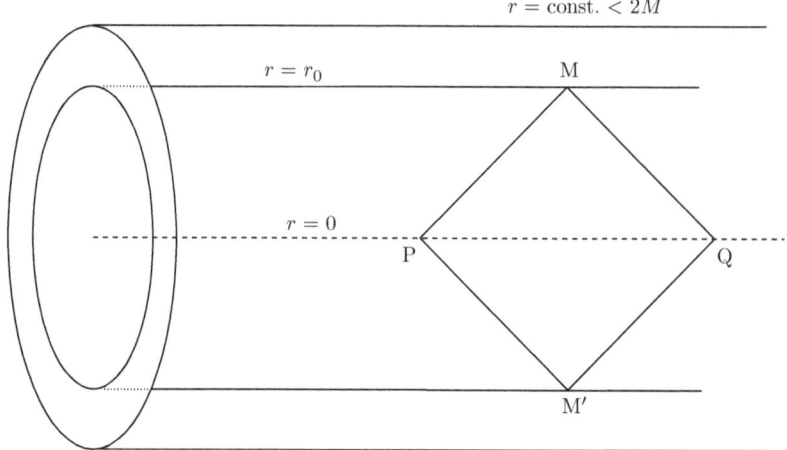

Fig. 5.5 Hypersurfaces $r = $ constant inside a Schwarzschild black hole with mass M; the lines MP, MQ and M'P, M'Q are null rays starting at $r = r_0 < 2M$ and reaching the singularity at $r = 0$.

$r = 0$ is time-like and there exists an inner apparent horizon which is at the same time a Cauchy horizon. It has been shown by Penrose (1968) that the inner horizon is a surface of infinite blueshift and is unstable when perturbed. This property lies at the origin of the phenomenon of mass inflation discovered by Poisson and Israel (1989, 1990) which we now briefly describe.

Because of the interaction with the space–time curvature (acting as a potential barrier) part of the gravitational radiation which is emitted during gravitational collapse is back-scattered onto the black hole and its energy density is infinitely blueshifted as it approaches the Cauchy horizon. The coexistence near the inner (Cauchy) horizon of the blueshifted, back-scattered radiation and of the outflux of radiation from the star as it shrinks within the black hole produces a spectacular effect; the gravitational mass and the curvature of space–time are inflated to values which are classically unlimited. This phenomenon has been dubbed *mass inflation*. Externally no trace of this is detectable as it happens inside the black hole. Outside observers continue to be influenced by a gravitational mass that is unchanged from the stellar precursor. From the point of view of space–time geometry the outflux of radiation merely plays the catalytic role of focusing generators which initiates the contraction of the generators of the Cauchy horizon. On the other hand, the large increase in the mass produces a deflation of the inner horizon and a separation of the Cauchy and inner horizons occurs. Mass inflation has the effect of making the charge and angular momentum (which are conserved) negligible. Hence the geometry near the Cauchy horizon becomes of Schwarzschild type and it is expected that a space-like singularity forms.

Provided one stays outside the region near the singularity where quantum gravity effects are important (i.e. a region where curvature becomes greater than $1/l_{\text{Pl}}^2$ where $l_{\text{Pl}} = \sqrt{\hbar G/c^3} = 10^{-33}$cm is the Planck length) it is legitimate to use the classical and even semiclassical laws of physics to explore the interior of a black hole. It is generally assumed that the problem of the existence of a singularity will be solved once gravity is quantized and space–time geometry is modified and replaced by a new singularity-free geometry. Following these ideas Markov (1984) proposed a limiting curvature principle. He claimed that the curvature invariant is bounded above so that $R_{ijkl}R^{ijkl} < 1/l^4$ where l is of the order of the Planck length l_{Pl}. For a Schwarzschild black hole this corresponds to values of the radial coordinate which are smaller than r_0 such that, in ordinary units,

$$R_{ijkl}R^{ijkl} = \frac{12(2GM/c^2)^2}{r_0^6} = \frac{1}{l_{\text{Pl}}^4}. \tag{5.153}$$

The value of r_0 for a stellar black hole is such that $l_{\text{Pl}} \ll r_0 \ll 2GM/c^2$. For values of r smaller than r_0 it is expected that vacuum polarization has a self-regulatory effect on the rise of curvature and that once the quantum fluctuations have died away the de Sitter state arises naturally (Markov and Mukhanov 1985, Mukhanov and Brandenberger 1992, Polchinski 1989, Israel and Poisson 1988). A model of a spontaneous transition from the Schwarzschild to the de Sitter metric along the space-like hypersurface $r = r_0$ was proposed by Frolov et al. (1990). We describe here another possibility based on the creation of disconnected de Sitter universes with light-like boundaries taking the form of light-like shells. Light-like or null shells are often used to provide simplified models for the description of gravitational collapse (see for instance Section 5.5 below). We start by presenting a summary of some of their basic properties.

5.4.1 Colliding null shells

A singular hypersurface in general relativity is defined as a hypersurface across which the metric is continuous but not its first derivatives. An immediate consequence of this is the presence of a singular term $\delta(\Phi(x))$ in the expression for the Riemann curvature where $\Phi(x) = 0$ is the equation of the singular hypersurface. When the hypersurface is time-like it represents the history of a thin shell and when the hypersurface is space-like it can be interpreted as a sudden phase transition, as in the model proposed by Frolov et al. (1990). In the light-like case we have in general coexistence of an impulsive gravitational wave and of a null shell (thin shell moving at the speed of light). The case of a pure impulsive wave (see Chapter 2) corresponds to a regular Ricci tensor and a Weyl tensor containing a singular Dirac δ-term, and for a pure null shell the Weyl tensor is regular but the Ricci tensor is singular and contains a Dirac δ-term. A general description of singular null hypersurfaces can be found in Barrabès and Hogan (2003b) and Barrabès and Israel (1991).

We limit ourselves here to spherically symmetric null shells. In terms of Eddington–Finkelstein retarded or advanced time w, the metric of a general spherically symmetric geometry is

$$ds^2 = -e^{\psi}\,dw(f\,e^{\psi}\,dw + 2\zeta\,dr) + r^2\,(d\theta^2 + \sin^2\theta\,d\phi^2), \tag{5.154}$$

where ψ, f are functions of w and r. The sign factor ζ is $+(-)1$ if r increases (decreases) toward the future along a ray $w = $ constant. Consider a thin shell whose history Σ is a light-cone $w = $ constant, which splits space–time into two domains $\mathcal{M}_\pm$, where $+(-)$ refers to the future(past) side of Σ. The metric in $\mathcal{M}_\pm$ has the form (5.154) with different functions $\psi_\pm$, $f_\pm$. We assume that ζ is the same on both sides of Σ (the case $\zeta_+ \neq \zeta_-$ occurring for instance when Σ is a common horizon to $\mathcal{M}_\pm$). The intrinsic metric on Σ is

$$ds^2|_\Sigma = r^2 \left(d\theta^2 + \sin^2\theta \, d\phi^2\right) = g_{ab}\, d\xi^a d\xi^b, \tag{5.155}$$

where $\xi^a = (r, \xi^A) = (r, \theta, \phi)$, with $a = 1, 2, 3$ and $A = 2, 3$, are intrinsic coordinates on Σ and the three basis vectors $e_{(a)} = \partial/\partial \xi^a$ are tangent to Σ. The metric g_{ab} is degenerate on a light-like hypersurface and one takes as pseudo-inverse the symmetric matrix g_*^{ab} formed by bordering with zeros the contravariant 2-metric g^{AB} inverse of g_{AB}. The normal to a light-like hypersurface is tangent to its null generators. It is convenient to choose as future-directed light-like vector n normal to Σ

$$n^a = \zeta \frac{\partial x^a}{\partial r}. \tag{5.156}$$

It can be shown that the surface stress–energy tensor $S^{ij} = S^{ab}\, e^i_{(a)} e^j_{(b)}$ of a spherically symmetric light-like shell having the null hypersurface Σ as history takes the perfect fluid form

$$S^{ab} = \mu n^a n^b + P g_*^{ab}, \tag{5.157}$$

where μ and P represent respectively the surface energy-density and surface pressure of the thin shell and are given by

$$\mu = \frac{\zeta}{8\pi r}(f_+ - f_-), \qquad P = -\frac{\zeta}{8\pi}(\partial_r \psi_+ - \partial_r \psi_-). \tag{5.158}$$

If one introduces the mass function $M(r, w)$, such that $f = 1 - 2M(r, w)/r$, then the mass of the shell, $m \equiv 4\pi r^2 \mu$, reads

$$m = -\zeta(M_+ - M_-), \tag{5.159}$$

and is generally a function of r and w. In what follows we assume that $\mathcal{M}_\pm$ have static, spherically symmetric geometry, such as Schwarzchild, Reissner–Nordström, de Sitter, or any superposition of these. Then there is no dependence on the time coordinate w and one can set $\psi = 0$. The parameter r is affine and the null shell is pressureless ($P = 0$).

Consider the collision of two spherical light-like shells, and in particular two concentric shells (one ingoing and the other outgoing) which collide in a 2-sphere S with radius r_0. The two shells re-emerge from S as two new light-like spherical shells. We call the corresponding null hyperfaces Σ_3 and Σ_4 before the collision, and Σ_1 and Σ_2 after the collision (see Figure 5.6). These null hypersurfaces divide space–time near S into four sectors which we label clockwise from noon as A, B, C, D. We introduce the index $I = 1, 2, 3, 4$ labelling the null shells. On each Σ_I we choose a parameter λ_I (not necessarily affine but the same on both faces) along the null generators of Σ_I and denote by $l^a_{(I)} \partial_a = \partial/\partial \lambda_I$ the light-like vector tangent to the generators (it is often convenient to take $\lambda_I = r$). The dilation rate K_I of an element of intrinsic 2-area

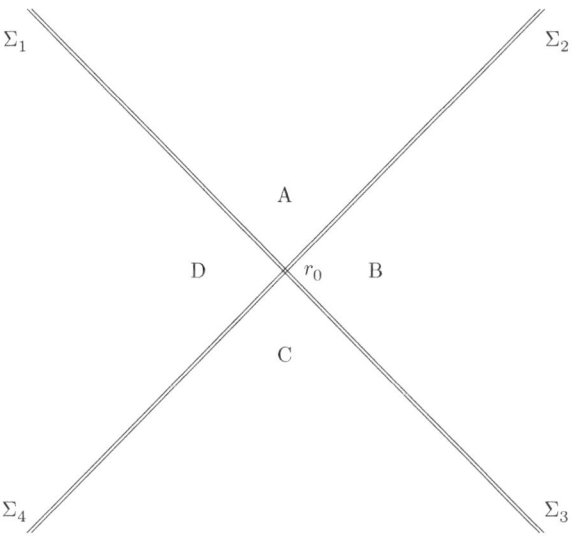

Fig. 5.6 Collision of two concentric, spherical, light-like shells Σ_3, Σ_4 at radius $r = r_0$ resulting in two concentric, spherical, light-like shells Σ_1, Σ_2.

convected along these generators is equal to $K_I = 2r^{-1} l_{(I)}^a \partial_a r$, and we form the four scalar functions

$$F_A = \frac{K_1 K_2}{l_{(1)} \cdot l_{(2)}}, \quad F_B = \frac{K_3 K_4}{l_{(3)} \cdot l_{(4)}}, \quad F_C = \frac{K_2 K_3}{l_{(2)} \cdot l_{(3)}}, \quad F_D = \frac{K_1 K_4}{l_{(1)} \cdot l_{(4)}}. \tag{5.160}$$

Using the property $(l_{(1)} \cdot l_{(2)})(l_{(3)} \cdot l_{(4)}) = (l_{(1)} \cdot l_{(4)})(l_{(2)} \cdot l_{(3)})$ one sees that, at each point of the 2-sphere of collision S,

$$F_A F_B = F_C F_D. \tag{5.161}$$

Introducing $f = g^{ab} r_{,a} r_{,b} = g^{rr}$ this relation becomes

$$f_A(r_0) f_B(r_0) = f_C(r_0) f_D(r_0). \tag{5.162}$$

It was first derived by Dray and 't Hooft (1985) and Redmount (1985) and later generalized to cases including rotating black holes by Barrabès et al. (1990). If one introduces the local mass functions M_A, M_B, M_C, M_D, then (5.162) yields a local relation between their values at the collision. For weak fields, to linear order (neglecting quadratic potential energy terms) it expresses conservation of gravitational mass in the collision. We note that a conservation relation also exists between the masses of the shells. Using (5.159) to define the masses m_I of the shells, and the sign convention for ζ, we find that

$$m_1 = M_A - M_D, \quad m_2 = M_C - M_A, \quad m_3 = M_C - M_B, \quad m_4 = M_B - M_D, \tag{5.163}$$

from which immediately follows the conservation relation $m_1 + m_2 = m_3 + m_4$ between the masses of the null shells.

5.4.2 Creation of de Sitter universes

In the model proposed by Frolov et al. (1990) and discussed above the singularity at $r = 0$ of the Schwarzschild geometry is eliminated by the occurence of an instantaneous phase transition from the Schwarzschild geometry to the de Sitter geometry along the whole space-like hypersurface $r = r_0$ inside the black hole [see (5.153)]. It seems however natural to expect that such a transition would occur randomly both in space and time. Hence we shall study the model, presented by Barrabès and Frolov (1996), of a spontaneous creation of de Sitter phase bubbles at arbitrary points along the hypersurface r_0. For the creation of a single bubble we use the properties of colliding null shells described above. We assume in Figure 5.6 that only the null shells Σ_1 and Σ_2 are present and therefore that the sectors B, C, and D are identical and not separated. The corresponding figure describes the spontaneous creation of a pair of light-like shells with sector A having de Sitter geometry and sectors B, C, D having the same Schwarzschild geometry. Since $f_B(r_0) = f_C(r_0) = f_D(r_0) \neq 0$ the relation (5.162) leads to

$$f_A(r_0) = f_B(r_0), \tag{5.164}$$

with $f_B(r) = 1 - 2M/r$, $f_A(r) = 1 - r^2/a^2$. The mass of the black hole is M and a is the radius of the de Sitter horizon so that $a^2 = 3/\Lambda = 3/8\pi\rho_{\rm dS}$. We deduce immediately the following relation

$$r_0^3 = 2M\,a^2, \tag{5.165}$$

giving $a \ll r_0 \ll 2M$. The creation of the de Sitter bubble occurs in the region of the de Sitter space–time where all future-directed light rays contract and both shells converge towards $r = 0$ (see Figure 5.7). Using (5.158) with $\zeta = -1$, and (5.165), the mass $m(r) = 4\pi r^2 \mu(r)$ of the two light-like shells takes the common value

$$m(r) = -M\left(1 - \frac{r^3}{r_0^3}\right). \tag{5.166}$$

This relation shows that $m(r)$ is zero at the moment of creation of the de Sitter phase bubble and that it is later negative and equal to $m(0) = -M$ when the shells hit the singularity $r = 0$. By comparison with the model proposed by Frolov et al. (1990), where the transition between the Schwarzschild and de Sitter space–times occurs instantaneously at $r = r_0$, we now have a situation which is no longer homogeneous as one moves along this hypersurface. This allows for the possibility of simultaneous creation of several de Sitter bubbles along $r = r_0$ which will enhance the inhomogeneity of this hypersurface. The Schwarzschild line-element near $r = 0$ can be approximated as

$$ds^2 \sim -\frac{r}{2M}\,dr^2 + \frac{r}{2M}\,dt^2 + r^2\,d\Omega^2. \tag{5.167}$$

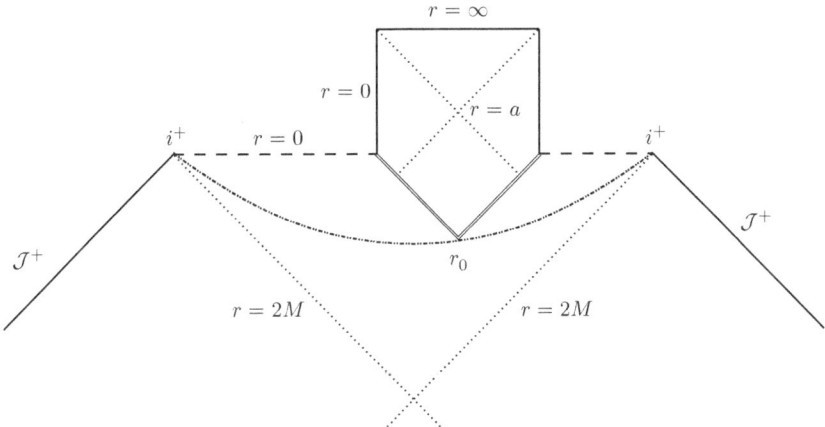

Fig. 5.7 Creation of a single de Sitter phase bubble, with parameter a, at $r = r_0$ within the horizon of a Schwarzschild black hole of mass M; the lower pair of dotted lines represent the Schwarzschild horizon $r = 2M$ and the upper pair represent the de Sitter horizon $r = a$ (Barrabès and Frolov, 1996).

Introducing the proper-time coordinate τ via $d\tau = -(r/2M)dr$ with $dr < 0$ we arrive at

$$ds^2 = -d\tau^2 + \left(-\frac{4M}{3\tau}\right)^{2/3} dt^2 + \left(\frac{9M\tau^2}{2}\right)^{2/3} d\Omega^2. \tag{5.168}$$

Consider a couple of de Sitter bubbles created at two different vertices both located on the hypersurface $r = r_0$. If the vertices are close enough their light-like boundaries may intersect and different scenarios of what can happen at the intersection can be proposed. It can be shown (Barrabès and Frolov, 1996) that the maximum coordinate distance between the vertices in order to have intersection is $\Delta t_{\max} = r_0^2/2M = a^2/r_0$. When the vertices are separated by a distance smaller than $\Delta t_{\max}$ the light-like boundaries intersect at some value r_1 of the radial coordinate such that $0 < r_1 < r_0$. Assume that the null shells cross each other without any interaction other than gravitational interaction, and that a new de Sitter space–time with horizon $a' \neq a$ forms in the future of r_1. Applying the general relation (5.162) for colliding null shells we arrive at two different possibilities. For $a' < r_1 < a$ the new de Sitter universe coexists indefinitely with the two initial ones until $r \to \infty$ (Figure 5.8a) and for $a < a' < r_1$ it finally occupies the whole space (Figure 5.8b). Another scenario in which the two null shells merge into a time-like shell separating the two initial de Sitter bubbles has been considered by Barrabès and Frolov (1996). All of these models show that a large number of disconnected de Sitter universes can be created within the time $t_{\text{evap}} = t_{\text{Pl}}(M/M_{\text{Pl}})^3$ needed for the quantum evaporation of the black hole. Such scenarios provide a classical singularity-free model of a black hole interior which might have interesting applications to the information-loss puzzle since it opens connections to other universes.

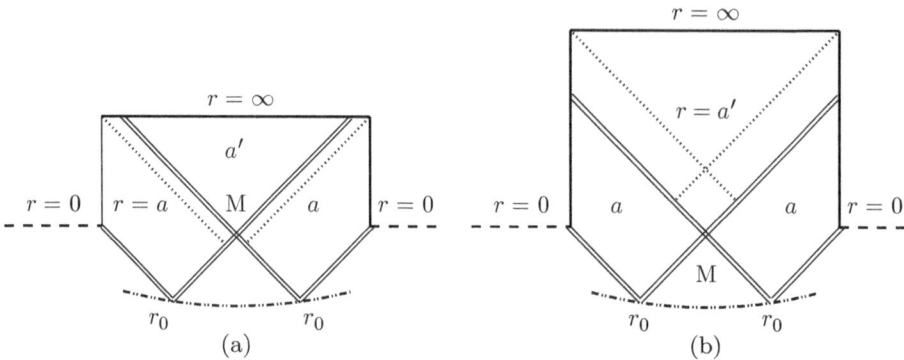

Fig. 5.8 Creation of a pair of de Sitter phase bubbles with parameter a at $r = r_0$ followed by the intersection of their boundaries at $M(r = r_1)$ and the subsequent production of a new de Sitter universe with parameter a'; case (a) corresponds to $a' < r_1 < a$ and case (b) to $a < a' < r_1$ (Barrabès and Frolov, 1996).

5.5 Metric fluctuations and Hawking radiation

In his original derivation of black hole radiance Hawking (1975) considered the propagation of a linear quantized field in a classical background geometry [see also Birrel and Davies (1982) and Carroll (2004)]. The mean value of the energy–momentum tensor is taken as the source of the gravitational field which is itself treated classically. The validity of this semiclassical approach has been questioned, in particular the controversial role of arbitrarily large ('transplanckian') frequencies of vacuum fluctuation ('t Hooft 1985, 1996, Brout et al. 1995) and the impact of gravitational back reaction due the emission of quanta. Zero-point fluctuations of quantum fields induce fluctuations of the metric. We study in this section how the fluctuations of the black hole geometry affect the properties of Hawking radiation (Barrabès et al., 1999). To describe these fluctuations quantum mechanically and to determine their effects on Hawking radiation requires full quantum gravity which is technically a very complicated problem. The modifications of Hawking radiation due to the metric fluctuations will be extracted within a simpler framework in which these fluctuations are treated classically. Furthermore only spherically symmetric fluctuations will be considered and the scattering by the gravitational potential barrier which occurs in the 4-dimensional d'Alembertian will be neglected.

We use a model proposed by York (1983) in which the fluctuating geometry near the horizon of the black hole is represented by a Vaidya metric with a fluctuating mass. Using an advanced time coordinate the Vaidya line-element reads

$$ds^2 = -\left(1 - \frac{2m(v)}{r}\right) dv^2 + 2dv\, dr + r^2\, d\Omega^2, \tag{5.169}$$

with here

$$m(v) = M\left[1 + \mu(v)\right]\theta(v), \tag{5.170}$$

Metric fluctuations and Hawking radiation

$$\mu(v) = \mu_0 \sin(\omega v)\theta(v). \tag{5.171}$$

The mass of the black hole fluctuates with frequency ω and the dimensionless amplitude μ_0 is assumed to be small ($\mu_0 \ll 1$) for a black hole with mass M much larger than the Planck mass. The Heaviside step function $\theta(v)$ in (5.170) indicates that the formation of the black hole results from the collapse of a massive null shell with mass M. The origin $v = 0$ of the advanced-time coordinate is taken to correspond to the collapse of the shell (see Figure 5.9). Hence for $v < 0$ the space–time geometry is flat while it is of Vaidya type when $v > 0$.

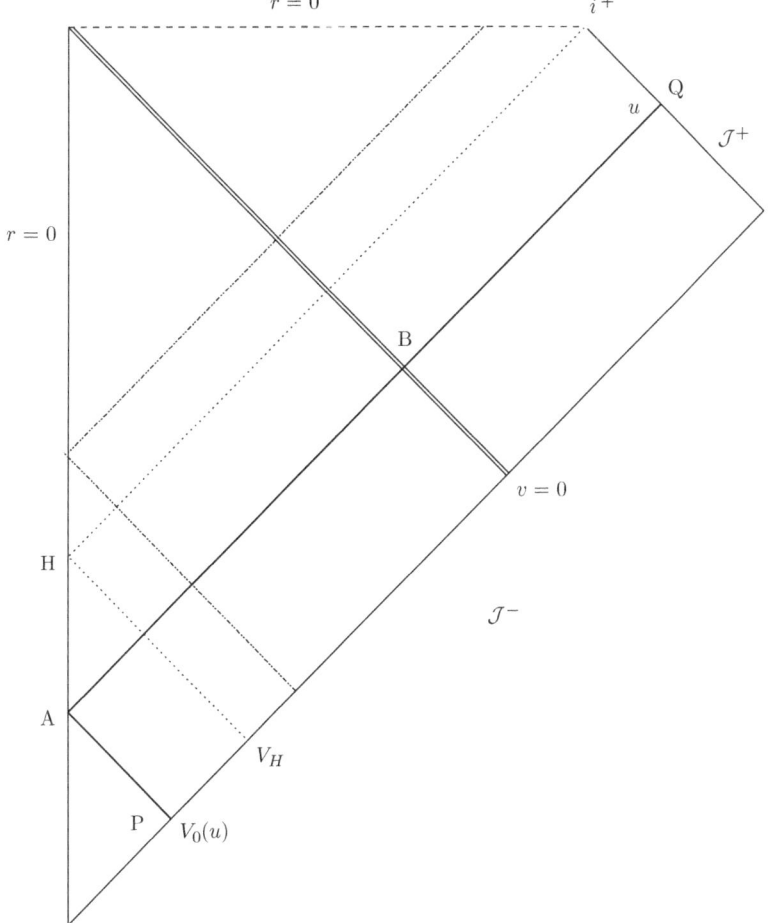

Fig. 5.9 Conformal diagram of a black hole formed by the collapse of a spherical null shell; the double line $v = 0$ is the history of the collapsing null shell. The solid dark line PABQ represents a light ray leaving $\mathcal{J}^-$ at advanced time $v = V_0(u)$ and reaching $\mathcal{J}^+$ at retarded time u (Barrabès et al., 1999).

The energy–momentum tensor associated with the metric (5.169) is given by

$$T_{ab} = \frac{l_a l_b}{4\pi r^2} \left[M\delta(v) + M\mu(v)\theta(v) \right], \tag{5.172}$$

where $l_a = -v_{,a}$ is a future-directed null vector field tangent to ingoing radial null rays. The first term in T_{ab} corresponds to the collapsing null shell whose space–time history is the null hypersurface $v = 0$ and the second term is the source of the Vaidya metric. We use the geometric optics approximation to obtain the solution of the wave equation and to study the propagation of radial null rays in the fluctuating geometry (5.169). Ingoing radial null rays are given by $v = $ constant, while outgoing null rays obey the equation

$$\left(1 - \frac{2m(v)}{r}\right) dv = dr. \tag{5.173}$$

In order to solve this equation we use a perturbation method and write

$$r(v) = R(v) + \rho(v) + \sigma(v) + \cdots. \tag{5.174}$$

Here $R(v)$ is the solution in the absence of fluctuations ($\mu_0 = 0$) while $\rho(v)$ and $\sigma(v)$ are first- and second-order terms, respectively, in the perturbation parameter μ_0. We find from (5.173) when $v > 0$ that

$$2\frac{dR}{dv} = 1 - \frac{2M}{R}, \tag{5.175}$$

$$2\frac{d\rho}{dv} - \frac{2M}{R^2}\rho = -\frac{2M}{R}\mu, \tag{5.176}$$

$$2\frac{d\sigma}{dv} - \frac{2M}{R^2}\sigma = \frac{2M}{R}\left[\frac{\rho\mu}{R} - \frac{\rho^2}{R^2}\right]. \tag{5.177}$$

We begin by looking for the perturbed position, $r_H = R_H + \rho_H + \sigma_H + \cdots$, of the event horizon. From the zeroth-order equation R is constant and equal to $R_H = 2M$. Introduction of this value into (5.176) and (5.177) and solving these equations leads to the following:

$$\rho_H = 2M\mu_0 \left[\frac{\Omega \cos(\omega v) + \sin(\omega v)}{1 + \Omega^2}\right], \tag{5.178}$$

$$\sigma_H = 2M\mu_0^2 \left[\frac{2\Omega^2(2 - \Omega^2)\cos(2\omega v) + \Omega(1 - 5\Omega^2)\sin(2\omega v)}{2(1 + \Omega^2)^2(1 + 4\Omega^2)}\right]. \tag{5.179}$$

We have introduced here the dimensionless frequency $\Omega = \omega/\kappa$ where $\kappa = (4M)^{-1}$ is the surface gravity of the unperturbed black hole with mass M. The resulting expression of r_H shows that the position of the event horizon of the perturbed black hole fluctuates. It follows that the surface area $\mathcal{A}(v) = 4\pi r_H^2$ of the event horizon and the surface gravity $\kappa = m(v)/r_H^2$ also fluctuate and have mean values equal to

$$\bar{\mathcal{A}} \sim 16\pi M^2 \left[1 + \frac{\mu_0^2}{2(1 + \Omega^2)}\right], \tag{5.180}$$

and
$$\bar{\kappa} \sim \kappa\left[1 + \frac{\mu_0^2}{2(1+\Omega^2)}\right], \tag{5.181}$$

respectively. Computation of the modified Hawking flux of radiation will later show that $\bar{\kappa}$ is identical to the renormalized surface gravity. The mean value of the Hawking temperature is $\bar{T}_H = \bar{\kappa}/2\pi$ (in units for which the Boltzman constant is unity) and the changes of area $\delta \mathcal{A} = \bar{\mathcal{A}} - \mathcal{A}$ and of temperature $\delta T_H = \bar{T}_H - T_H$ obey the relation

$$\frac{\delta \mathcal{A}}{\mathcal{A}} = \frac{\delta T_H}{T_H}. \tag{5.182}$$

This induces a modification of the entropy of the black hole. Its mean value satisfies $d\bar{S} = dE/\bar{T}_H$ where $E = \bar{m}(v) = M$ and it is equal to

$$\bar{S} = \frac{\bar{\mathcal{A}}}{4}\left(1 - \frac{\mu_0^2}{2(1+\Omega^2)}\right), \tag{5.183}$$

instead of the usual relation $S = \mathcal{A}/4$.

We start by studying Hawking radiation in the absence of metric fluctuations (so that $\mu_0 = 0$) and considering an outgoing null ray which reaches future null infinity $\mathcal{J}^+$ at a given value u of retarded time. Tracing back in time this null ray one arrives at the trajectory PABQ on Figure 5.9. Such a ray starts from past null infinity $\mathcal{J}^-$, converges radially and diverges again after bouncing off at ($r = 0$). On its way towards $\mathcal{J}^+$ it crosses the null shell where $v = 0$. Our aim is to obtain the relation $V_0(u)$ between the value of the advanced time at the departure of the ray from $\mathcal{J}^-$ and the value of the retarded time at its arrival at $\mathcal{J}^+$. The future of a null ray leaving $\mathcal{J}^-$ has a different destination depending upon the value of the advanced time at its departure. Looking at Figure 5.9 and considering all possibilities for ingoing null rays leaving $\mathcal{J}^-$ at different values of v one sees that: (i) for $v > 0$ the ray directly hits the singularity $r = 0$; (ii) for $v < V_H$ the ray reaches $\mathcal{J}^+$ after reflection at $r = 0$; and (iii) for $V_H < v < 0$ the ray propagates in the trapped region until reaching the singularity after reflection at $r = 0$. Only case (ii) is relevant for the study of Hawking radiation.

We first derive the relation $V_0(u)$ in the absence of fluctuation ($\mu_0 = 0$). In the flat space–time domain ($v < 0$) we have the usual relation $v - u = 2r$ and in the Schwarzschild domain ($v > 0$) this relation becomes $v - u = 2[r - 2M + 2M \ln\{(r - 2M)/2M\}]$ [see (5.5) and (5.6)]. In both domains v (u) is constant along an ingoing (outgoing) radial null ray. Calling V_0 the initial value of advanced time at departure from $\mathcal{J}^-$, and R_0 the value of the radial coordinate when the ray crosses the shell at $v = 0$, it can be shown that

$$V_0 = -2R_0 \quad \text{and} \quad -u = 2\left[R_0 - 2M + 2M \ln\left(\frac{R_0}{2M} - 1\right)\right]. \tag{5.184}$$

The first relation in (5.184) is obtained when $v < 0$ and the second when $v > 0$. Since the radial coordinate is continuous across the singular null hypersurface ($v = 0$) these relations combine to give the required expression for $V_0(u)$.

In the presence of metric fluctuations ($\mu_0 \neq 0$) the part of the trajectory in the domain $v > 0$ is modified and to the same retarded time u at arrival at $\mathcal{J}^+$ will correspond a new value $V(u)$ of advanced time at departure from $\mathcal{J}^-$. The first equation in (5.184) is now replaced by

$$V(u) = -2[R_0(u) + \rho_0(u) + \sigma_0(u)], \tag{5.185}$$

where the subscript zero still refers to the interesection of the ray with the hypersurface $v = 0$. The first- and second-order terms ρ and σ of (5.174) are solutions of the equations (5.176) and (5.177). These equations can be rewritten in the common form

$$\frac{df}{dv} - Mf^2 = F, \tag{5.186}$$

with

$$f = \rho, \qquad F = -\frac{M}{R}, \tag{5.187}$$

for the first-order perturbation and

$$f = \sigma, \qquad F = \frac{M}{R^2}\mu\rho - \frac{M}{R^3}\rho^2, \tag{5.188}$$

for the second-order perturbation. It is convenient to use (5.175) in order to replace the variable v by the dimensionless variable $x = (R-2M)/2M$ and to introduce the dimensionless quantities

$$\tilde{u} = \kappa u, \qquad \tilde{V} = \kappa V, \qquad \tilde{\rho} = \frac{\rho}{2M}, \qquad \tilde{\sigma} = \frac{\sigma}{2M}. \tag{5.189}$$

We then find that

$$\tilde{\rho}(x) = \frac{x}{1+x} \int_x^\infty \frac{d\xi}{\xi^2}(1+\xi)\hat{\mu}(\xi), \tag{5.190}$$

$$\tilde{\sigma}(x) = -\frac{x}{1+x} \int_x^\infty \frac{d\xi}{\xi^2}\tilde{\rho}(\xi)\left[\hat{\mu}(\xi) - \frac{\tilde{\rho}(\xi)}{1+\xi}\right], \tag{5.191}$$

where $\hat{\mu}(\xi) = \mu_0 \sin[\Omega(\xi + \ln\xi + \tilde{u})]$. In these equations the constant of integration has been chosen in such a way that the perturbed ray arrives at $\mathcal{J}^+$ at the same retarded time u as the unperturbed one. Introducing these results into (5.185) results in

$$\tilde{V}(\tilde{u}) = -[1 + x_0 + \tilde{\rho}(x_0) + \tilde{\sigma}(x_0)], \tag{5.192}$$

where $x_0 = (R_0 - 2M)/2M$ satisfies

$$\tilde{u} = -x_0 + \ln x_0. \tag{5.193}$$

We are interested in the rays propagating near the horizon and arriving at $\mathcal{J}^+$ at late time $u \gg 1$. Solving (5.193) by iteration gives

$$x_0 = e^{-\tilde{u}}(1 - e^{-\tilde{u}}) + O(e^{-3\tilde{u}}), \tag{5.194}$$

showing that x_0 is a very small quantity. After some lengthy calculations it can be shown that $\tilde{V}(\tilde{u})$ takes the following form [cf. equations (5.2)–(5.5) of Barrabès et al. (1999)]

$$\tilde{V}(\tilde{u}) = -e^{-\tilde{u}}[1 + A_1 \sin(\Omega\tilde{u} + \varphi_1) + A_2 \sin(2\Omega\tilde{u} + \varphi_2) + C\tilde{u}], \tag{5.195}$$

where A_1, A_2 and the phases φ_1, φ_2 are functions of the dimensionless frequency $\Omega = \omega/\kappa$ and C is given by

$$C = -\frac{\mu_0^2}{2(1+\Omega^2)}. \tag{5.196}$$

We do not need A_2 or φ_2 for our final results. For A_1 and φ_1 we have

$$A_1 = \mu_0 \frac{q(\Omega)}{(1+\Omega^2)^{1/2}} + 2\mu_0^2 \frac{q(\Omega)(\Omega^2-1)}{\Omega(1+\Omega^2)^{3/2}}, \tag{5.197}$$

$$\varphi_1 = -\varphi_\Gamma(\Omega) + \arctan \Omega, \tag{5.198}$$

where $\varphi_\Gamma(\Omega)$ is a real function which, for large values of Ω, behaves as $\varphi_\Gamma \sim \Omega$, and

$$q(\Omega) = \frac{\sqrt{2\pi}}{\sqrt{\Omega(e^{2\pi\Omega}-1)}}. \tag{5.199}$$

Finally we study the energy flux and the asymptotic spectrum of Hawking radiation. To this end we calculate the contribution of the s-mode of a quantum scalar massless field to Hawking radiation. Neglecting the scattering by the gravitational potential barrier amounts to using a 2-dimensional approximation in which ingoing and outgoing modes decouple completely. Assuming that the field is in its vacuum state before the formation of the black hole, we use the following relation which gives the mean energy flux of Hawking radiation at $\mathcal{J}^+$:

$$\frac{dE}{du} \equiv 4\pi r^2 \langle T_{uu}\rangle^{\text{ren}} = \frac{\kappa^2}{12\pi}\left(\frac{d\tilde{V}}{d\tilde{u}}\right)^{1/2} \frac{d^2}{d\tilde{u}^2}\left[\left(\frac{d\tilde{V}}{d\tilde{u}}\right)^{-1/2}\right]. \tag{5.200}$$

Introducing (5.195) into this equation produces an expression which can be written as the sum of a permanent part and a fluctuating part. Only the permanent part will contribute to the total energy received at $\mathcal{J}^+$. Direct calculation yields

$$\left(\frac{dE}{du}\right)^{\text{perm}} = \frac{\kappa^2}{48\pi}\left[1 + \frac{1}{2}\mu_0^2 \Omega^2 q^2(\Omega) - 2C\right], \tag{5.201}$$

with C given by (5.196). This latter term, which gave a linear contribution in (5.195), does not contribute to the fluctuating part of dE/du. Since $\mu_0 \ll 1$ it can be absorbed into $e^{\tilde{u}}$ and removed from (5.195). Such a transformation corresponds to the renormalization of the surface gravity, $\kappa \to \kappa_r = \kappa(1-C)$. We note that it is identical to the mean value $\bar{\kappa}$ introduced earlier in (5.181). Hence the permanent part of the mean energy flux at $\mathcal{J}^+$ can be rewritten as

$$\left(\frac{dE}{du}\right)^{\text{perm}} = \frac{\kappa_r^2}{48\pi}\left[1 + \mu_0^2 \frac{\pi\Omega}{e^{2\pi\Omega}-1}\right]. \tag{5.202}$$

The modifications of the energy flux, which are due to the fluctuations of the metric, manifest themselves in the renormalized surface gravity [see (5.196)] and in the additional term $\mu_0^2 \pi\Omega/(e^{2\pi\Omega}-1)$. Both modifications are of second order in μ_0 and decrease when the frequency of the fluctuations increases. In the absence of fluctuations the usual expression of $(dE/du)_0 = \kappa^2/(48\pi)$ holds.

Still using the 2-dimensional aproximation, and neglecting the effect of the gravitational barrier, we can extract the asymptotic spectrum of Hawking radiation from the properties of the Bogoliubov coefficients (Birrel and Davies 1982, Carroll 2004, Brout et al. 1995). These coefficients are given by the overlap of the initial (ingoing) modes which are specified at $\mathcal{J}^-$ and the final (outgoing) modes specified at $\mathcal{J}^+$. Since s-modes satisfy the 2-dimensional equation $\partial_u \partial_v \phi = 0$ they can be decomposed in terms of plane waves as

$$\phi_\nu(v) = \frac{e^{-i\nu v}}{\sqrt{4\pi\nu}}, \qquad \phi_\lambda(u) = \frac{e^{-i\lambda u}}{\sqrt{4\pi\lambda}}, \tag{5.203}$$

where ν is the energy measured at $\mathcal{J}^-$ for the in-modes, and λ is the energy measured at $\mathcal{J}^+$ for the out-modes. The reflection condition at $r = 0$ implies that the scattered in-modes are given by $\phi_\nu(V(u))$. Then the Bogoliubov coefficients are given by

$$\alpha_{\nu,\lambda} = \int du\, \phi_\nu^*(V(u))\, i \overleftrightarrow{\partial_u} \phi_\lambda(u) = \int du\, \frac{e^{i\nu V(u)}}{\sqrt{4\pi\nu}} \frac{e^{-i\lambda u}}{\sqrt{\pi\lambda^{-1}}}, \tag{5.204}$$

$$\beta_{\nu,\lambda} = \int du\, \phi_\nu(V(u))\, i \overleftrightarrow{\partial_u} \phi_\lambda(u) = \int du\, \frac{e^{-i\nu V(u)}}{\sqrt{4\pi\nu}} \frac{e^{-i\lambda u}}{\sqrt{\pi\lambda^{-1}}}. \tag{5.205}$$

In the unperturbed geometry one sees from (5.192) and (5.193) that for large u, $\kappa V_0(u) = -1 - e^{-\kappa u}$ approximately. Extending the domain of validity of the asymptotic behaviour of $V_0(u)$ for all u, one finds that

$$\alpha_{\nu,\lambda} = B(\nu,\lambda) \frac{e^{\pi\lambda/\kappa} e^{-\nu/\kappa}}{\sqrt{2\pi\kappa\nu}}, \qquad \beta_{\nu,\lambda} = B(\nu,\lambda) \frac{e^{\pi\lambda/\kappa}}{\sqrt{2\pi\kappa\nu}}, \tag{5.206}$$

where the function $B(\nu,\lambda)$ is given by

$$B(\nu,\lambda) = \Gamma\left(\frac{i\lambda}{\kappa}\right) \sqrt{\frac{\lambda}{2\pi\kappa}} \left(\frac{\nu}{\kappa}\right)^{-i\lambda/\kappa} e^{-\pi\lambda/2\kappa}, \tag{5.207}$$

and $\Gamma(z)$ is the gamma function. The mean number of quanta reaching $\mathcal{J}^+$ per unit retarded time is derived from the general expression

$$<\bar{n}_\lambda> = \frac{\int^N d\nu\, |\beta_{\nu,\lambda}|^2}{\int^N d\nu/\kappa\nu}. \tag{5.208}$$

The time average here is obtained by integrating over ν up to the cut-off frequency N and then dividing the resulting expression by the denominator as indicated.

Without fluctuations of the metric the mean number of quanta reaching $\mathcal{J}^+$ takes the form

$$<\bar{n}_\lambda>_0 = \frac{1}{2\pi} \frac{1}{e^{2\pi\lambda/\kappa} - 1}. \tag{5.209}$$

This is a Planck distribution with temperature $T = \kappa/2\pi = (8\pi M)^{-1}$. In the presence of fluctuations one needs to use the perturbed expression (5.195) for $V(u)$.

This leads to a modification of the mean rate of quanta reaching $\mathcal{J}^+$ which is now equal to

$$\langle \bar{n}_\lambda \rangle = \frac{1}{2\pi} \frac{1}{e^{2\pi\lambda/\kappa_r} - 1} - \left(\frac{A_1}{2\kappa_r}\right)^2 \frac{2\lambda}{e^{2\pi\lambda/\kappa_r} - 1},$$

$$+ \left(\frac{A_1}{2\kappa_r}\right)^2 \left[\frac{\lambda - \omega}{e^{2\pi(\lambda-\omega)/\kappa_r} - 1} + \frac{\lambda + \omega}{e^{2\pi(\lambda+\omega)/\kappa_r} - 1}\right], \tag{5.210}$$

where A_1 is given by (5.197). The modifications of the asymptotic spectrum of Hawking radiation appear in the renormalization of surface gravity and in the existence of three additional terms. Among them the two last terms in (5.210) contain Bose thermal factors showing that $\pm \omega$ plays the role of a chemical potential which facilitates leakage of energy and is reminiscent of superradiance.

6
Higher dimensional black holes

Higher dimensional space–times, with dimensions $D = 4 + k$ with $k > 0$, are a common ingredient in most theories attempting to unify gravity with the other forces of nature. This idea, which originated a long time ago with the work of Kaluza (1921) and Klein (1926), has received a new impetus from later developments in field theory and, in particular, string theories. The k extra dimensions have so far eluded detection, the conventional explanation being that they are compactified within radii of the order of the Planck length $l_{Pl} \sim 10^{-33}$cm. This corresponds to energies of the order of $E_{Pl} = 10^{19}$ GeV which are well out of reach compared to energies currently available. Some recent models consider larger dimensions [as large as a millimetre (Antoniadis et al., 1998)] and even infinite extra dimensions (Randall and Sundrum, 1999). They have received a lot of attention because they offer a way to lower energy from the Planck scale to the weak scale (of TeV order) and to provide an explanation for the large disparity between these two scales (the so-called hierarchy problem). They also raise the prospect of experimental verification and, in particular, the production of mini black holes at future colliders. In the so-called brane-world models the standard fields are confined to a 4-dimensional time-like hypersurface (the brane) embedded in a higher dimensional space–time (the bulk) where only gravity can propagate. Black holes on the other hand can be either attached to the brane or move in the bulk.

Let us assume that the k extra dimensions have radius of order l, and call respectively $M_{Pl,D}$ and G_D the Planck mass and the gravitational coupling constant for a D-dimensional space–time. In the ordinary case $D = 4$ the D index is omitted and we write simply M_{Pl} and G for these quantities. At distances $r \ll l$ much smaller than the radii of the extra dimensions the gravitational force has the usual expression derived from Gauss' law:

$$F = G_D \frac{m_1 m_2}{r^{2+k}}. \tag{6.1}$$

On the other hand, when $r \gg l$ the extra dimensions can be ignored and the gravitational force can be approximated as

$$F \sim G_D \frac{m_1 m_2}{l^k r^2}. \tag{6.2}$$

Hence the corresponding effective 4-dimensional gravitational constant G is such that $G \sim G_D l^{-k}$, in units for which $\hbar = c = 1$. From the definition of the Planck mass in arbitrary dimensions one has $G_D M_{Pl,D}^{D-2} = 1$ and therefore

$$M_{Pl,D}^{k+2} \sim M_{Pl}^2 \, l^{-k}. \tag{6.3}$$

This relation shows that increasing the size of the extra dimensions lowers the Planck scale which can eventually be made of the order of the electroweak scale $M_{EW} \sim$ TeV (for example with $k = 2$ and $l \sim 100$ µm–1 mm).

Another interesting consequence of large extra dimensions is the possible production of mini black holes at an energy scale which is accessible to high-energy experiments and exists in cosmic rays (Eardley and Giddings 2002, Rychkov 2004, Yoshino and Rychkov 2005, Cardoso et al. 2005, Horowitz 2012). Describing scattering processes of elementary particles leading to the production of black holes requires a full theory of quantum gravity. Since in these processes the centre-of-mass energy is much larger than the Planck mass a semiclassical approach can be followed. It is thus important to extend to higher dimensions properties of black holes which are already known in four dimensions. In this chapter, after a brief description of $D > 4$-dimensional black holes, we generalize to arbitrary dimensions the geometrical inequalities, already presented in Chapter 5, which give conditions for the formation of a trapped surface. This has application to the formation of apparent horizons (and thus of a black hole) in the collision of high-energy particles, as in the work of Yoshino and Nambu (2002). Also since scattering by ultra-relativistic spinning particles requires a knowledge of their gravitational fields the metric of a D-dimensional Kerr black hole boosted to the speed of light will be derived. This has been used to propose a model of the gravitational field of a spinning radiation beam pulse (a gyraton) in higher dimensions (Frolov and Fursaev 2005, Frolov, Israel and Zelnikov 2005).

6.1 Brief outline of *D*-dimensional black holes

Black holes are characterized by the presence of an event horizon. In a 4-dimensional stationary space–time a theorem of Hawking states that the surface topology of the event horizon has to be a 2-sphere, and furthermore uniqueness theorems for black hole solutions have been demonstrated. These properties are lost in dimensions larger than four. There exist solutions for which the event horizon has the topology of a $(D-2)$-sphere (S^{D-2}) and one still speaks of black holes. There also exist solutions with different topology, such as for example black rings where the event horizon has the topology $S^1 \times S^{D-3}$ (Emparan and Reall 2002, Emparan and Myers 2003, 2008).

The higher dimensional analogue of the Schwarzschild 4-D solution of the vacuum Einstein equations was obtained a long time ago by Tangherlini (1963). Its line-element has the form

$$ds^2 = -f(r)\,dt^2 + \frac{1}{f(r)}\,dr^2 + r^2\,d\Omega^2_{D-2}, \tag{6.4}$$

where r is the radial coordinate, and $d\Omega^2_{D-2}$ is the line-element of the unit $(D-2)$-sphere ($d\Omega^2_{n+1} = d\theta^2_{n+1} + \sin^2\theta_n\,d\Omega^2_n$ for $n \geq 1$). The function $f(r)$ is given by

$$f(r) = 1 - \frac{16\pi\,G_D\,M}{(D-2)s_{D-2}\,r^{D-3}}. \tag{6.5}$$

Here M is the mass of the black hole and s_n is the area of the unit sphere S^n. Thus

$$s_{D-2} = \frac{2\pi^{(D-1)/2}}{\Gamma(\frac{D-1}{2})}, \qquad (6.6)$$

with the gamma function appearing in the denominator. The Tangherlini solution has the same causal structure as the Schwarzschild solution. It has an event horizon corresponding to the value $r = r_H$ of the radial coordinate such that

$$r_H = \left(\frac{16\pi G_D M}{(D-2) s_{D-2}}\right)^{1/(D-3)}. \qquad (6.7)$$

Note that this relation shows that $G_D M \sim (\text{length})^{D-3}$.

The metric of a D-dimensional rotating black hole has been obtained by Myers and Perry (1986). While in four dimensions Kerr black holes have only two paramaters, the mass M and the angular momemtum parameter a, in $D > 4$ dimensions they are characterized by the mass M and l angular momentum parameters where l is given by

$$l = \left[\frac{D-1}{2}\right]. \qquad (6.8)$$

The notation $[N]$ indicates the integer part of N. Thus $l = (D-1)/2$ when D is odd and $l = (D-2)/2$ when D is even. The number l of angular momentum parameters is related to the existence of $[(D-1)/2]$ Casimirs of the group $SO(D-1)$ of spatial rotations. It represents here the number of independent 2-planes of rotation. The l angular momentum parameters will be denoted by a_i with $i = 1, 2, \ldots, l$ and the angular momentum in the i-th plane of rotation is

$$J_i = \frac{2M}{D-2} a_i. \qquad (6.9)$$

The Myers–Perry metric admits the $(l+1)$ Killing vectors $\xi_{(t)} = \partial_t$ and $\xi_{(i)} = \partial_{\phi_i}$ with $i = 1, 2, \ldots, l$. It is the most general metric describing the field of a D-dimensional rotating black hole and is most easily written in the Kerr-Schild form:

$$g_{ab} = \eta_{ab} + H\, k_a k_b, \qquad (6.10)$$

where k_a is a null vector and H a function of the coordinates x^a, with $a = 1, 2, \ldots, D$ and $x^D = t$ being the time cordinate. Also η_{ab} is the D-dimensional Minkowski metric, $k^a = g^{ab} k_b = \eta^{ab} k_b$ and $g^{ab} = \eta^{ab} - H k^a k^b$. For the spatial coordinates it is convenient to replace the set $(x^1, \ldots, x^{D-1})$ by either the $l = (D-1)/2$ pairs of coordinates (x^i, y^i) with $i = 1, 2, \ldots, l$ when D is odd or, when D is even, by the set $(x^i, y^i, x^{D-1} \equiv z)$ with now $l = (D-2)/2$. For odd values of D we have $l = (D-1)/2$ and we define

$$F = 1 - \sum_{i=1}^{l} \frac{a_i^2 \left((x^i)^2 + (y^i)^2\right)}{(r^2 + a_i^2)^2}, \qquad (6.11)$$

$$\Pi = \prod_{i=1}^{l} (r^2 + a_i^2), \qquad (6.12)$$

with
$$\sum_{i=1}^{l} \frac{(x^i)^2 + (y^i)^2}{r^2 + a_i^2} = 1. \tag{6.13}$$

The last equation implicitly defines r in terms of the spatial coordinates (x^i, y^i). The function H and the null vector k_a of the Myers–Perry metric (6.10) are given by
$$H = \frac{r_H^{D-3} r^2}{F \Pi}, \tag{6.14}$$

$$k_a \, dx^a = \sum_{i=1}^{l} \frac{r(x^i \, dx^i + y^i \, dy^i) + a_i(x^i \, dy^i - y^i \, dx^i)}{r^2 + a_i^2} - dt, \tag{6.15}$$

where r_H is the same as in (6.7). We note that r_H is not a horizon for a rotating black hole. When D is even F and Π continue to be defined as above but with now $l = (D-2)/2$ and r defined by
$$\sum_{i=1}^{l} \frac{(x^i)^2 + (y^i)^2}{r^2 + a_i^2} + \frac{z^2}{r^2} = 1. \tag{6.16}$$

The function H and the null vector k_a are now
$$H = \frac{r_H^{D-3} r}{F \Pi}, \tag{6.17}$$

$$k_a \, dx^a = \sum_{i=l}^{l} \frac{r(x^i \, dx^i + y^i \, dy^i) + a_i(x^i \, dy^i - y^i \, dx^i)}{r^2 + a_i^2} + \frac{z \, dz}{r} - dt. \tag{6.18}$$

When all of the angular parameters vanish we obtain the Kerr–Schild form of the Tangherlini metric.

The metric (6.10) can be put in Boyer–Lindquist form by considering r in (6.13) or (6.16) as a coordinate and introducing the angular coordinates μ_i, ϕ_i when D is odd, and μ_i, ϕ_i, α when D is even, defined by
$$x^i = (r^2 + a_i^2)^{1/2} \mu_i \cos\left(\phi_i - \tan^{-1} \frac{a_i}{r}\right), \tag{6.19}$$
$$y^i = (r^2 + a_i^2)^{1/2} \mu_i \sin\left(\phi_i - \tan^{-1} \frac{a_i}{r}\right), \tag{6.20}$$
$$z = \alpha \, r. \tag{6.21}$$

The relations (6.13) and (6.16) imply that when D is odd,
$$\sum_{i=1}^{l} \mu_i^2 = 1, \tag{6.22}$$
and when D is even,
$$\sum_{i=1}^{l} \mu_i^2 + \alpha^2 = 1. \tag{6.23}$$

Following further coordinate transformations $t \to \bar{t}$ and $\phi_i \to \bar{\phi}_i$, to eliminate some of the off-diagonal components of the metric, the line-element when D is odd becomes

$$ds^2 = \frac{\Pi F}{\Pi - r_H^{D-3} r^2} dr^2 + \sum_{i=1}^{l} (r^2 + a_i^2)(d\mu_i^2 + \mu_i^2 d\bar{\phi}_i^2)$$

$$+ \frac{r_H^{D-3} r^2}{\Pi F} (d\bar{t} + a_i \mu_i^2 d\bar{\phi}_i)^2 - d\bar{t}^2, \qquad (6.24)$$

and when D is even becomes

$$ds^2 = \frac{\Pi F}{\Pi - r_H^{D-3} r} dr^2 + \sum_{i=1}^{l} (r^2 + a_i^2)(d\mu_i^2 + \mu_i^2 d\bar{\phi}_i^2)$$

$$+ r^2 d\alpha^2 + \frac{r_H^{D-3} r}{\Pi F} (d\bar{t} + a_i \mu_i^2 d\bar{\phi}_i)^2 - d\bar{t}^2, \qquad (6.25)$$

with F and Π given by (6.11) and (6.12). Using these forms of the line-element for large values of r the interpretation of M, introduced in (6.5), and of J_i, introduced in (6.9), as, respectively, the mass and the angular momentum in the i-th plane of rotation is confirmed.

The horizons of a rotating D-dimensional black hole are given by the solutions of $g^{rr} = (g_{rr})^{-1} = 0$. From (6.24) and (6.25) these equations become

$$\Pi - r_H^{D-3} r^2 = 0, \qquad (6.26)$$

when D is odd and

$$\Pi - r_H^{D-3} r = 0, \qquad (6.27)$$

when D is even. The singularities, horizons, and their topology have been discussed by Myers and Perry (1986). Other properties concerning the existence of Killing–Yano tensors and symmetric Killing tensors generating symmetries which are analogous to those of the 4-dimensional Kerr black hole are described by Frolov and Sojkovic (2003), Frolov (2003), Frolov and Kubiznak (2007), and Frolov and Zelnikov (2011).

6.2 Gibbons–Penrose isoperimetric inequality and the hoop conjecture in D dimensions

In this section we generalize the results of the previous chapter to space–times with dimension $D > 4$. As in Chapter 5 we use the Penrose model of a convex thin shell collapsing in a vacuum from infinity with the speed of light. It is easy to see that the following equation still holds:

$$K = 16\pi G_D \mu, \qquad (6.28)$$

where K is the extrinsic curvature of the $(D-2)$-surface of the shell and μ is its surface energy density. The gravitational constant G_D now appears in (6.28) as this relation is obtained from the D-dimensional analogue of Raychaudhuri's equation (5.36). The total mean curvature Q of the shell is defined by

$$(D-2)\,Q = \int K dS. \tag{6.29}$$

With $M = \int \mu dS$ the total conserved mass (identical with the ADM mass and Bondi advanced mass) of the shell, we find from (6.28) and (6.29) that

$$(D-2)\,Q = 16\pi\, G_D\, M. \tag{6.30}$$

In arbitrary dimensions the Minkowski inequality (Minkowski, 1903) [see also Burago and Zalgaller (1988), p. 212] states that: *for any closed convex m-dimensional surface immersed in $\mathbf{R}^n$, with $2 \leq m < n$, its area $\mathcal{A}_m$ satisfies*

$$s_m\,(\mathcal{A}_m)^{m-1} \leq Q^m, \tag{6.31}$$

where Q is the total mean curvature and s_m is the area of the unit m-sphere [see (6.6)]. If (6.31) is applied to the imploding shell we obtain, since $m = D - 2$ and using (6.30), the D-dimensional form of the Gibbons–Penrose isoperimetric inequality

$$s_{D-2}\,(\mathcal{A}_{D-2})^{D-3} \leq \left(\frac{16\pi\, G_D\, M}{D-2}\right)^{D-2}. \tag{6.32}$$

According to this a marginally trapped surface, and therefore an apparent horizon, forms during the collapse of the shell provided its mass M is concentrated within a sufficiently compact domain whose boundary has an area satisfying (6.32). The Gibbons–Penrose inequality of general relativity, $\mathcal{A}_2 \leq 4\,(2GM)^2$, is recovered by putting $D = 4$ in (6.32). Introducing the Schwarzschild radius

$$r_H = \left(\frac{16\pi\, G_D\, M}{(D-2)\, s_{D-2}}\right)^{1/(D-3)}, \tag{6.33}$$

into (6.32) yields another familiar form of the Gibbons–Penrose inequality

$$(\mathcal{A}_{D-2})^{D-3} \leq s_{D-2}\, r_H^{D-2}. \tag{6.34}$$

We now consider the hoop conjecture in the present context. In order to generalize the inequalities (5.41) to higher dimensions we first need to obtain the generalized form of the geometrical inequalities (5.40). The following proposition has been established by Barrabès et al. (2004): *Let $\mathcal{D}$ be a convex domain of $\mathbf{R}^n$ and Q the total mean curvature of the boundary $\partial \mathcal{D}$ of $\mathcal{D}$. Let ω_{n-2} be the maximum area of the boundary of its orthogonal hyperplane projections, and Ω_{n-2} the maximum area of $(n-2)$-dimensional sections of $\partial \mathcal{D}$ by hyperplanes. Then the total mean curvature Q satisfies the following inequalities*

$$\frac{s_n}{2s_{n-2}}\,\Omega_{n-2} \leq Q \leq \frac{s_{n-1}}{s_{n-2}}\,\omega_{n-2}. \tag{6.35}$$

The derivation of this proposition is based upon two geometrical results. The first is the Cauchy formula which states that between the area $\mathcal{A}_{n-1}(\partial \mathcal{D})$ of the boundary

$\partial \mathcal{D}$ of a convex domain $\mathcal{D} \in \mathbf{R}^n$ and the mean volume $V_{n-1}(p_\xi(\mathcal{D}))$ of the orthogonal plane projections of $\mathcal{D}$ in an arbitrary direction there exists the relation

$$\mathcal{A}_{n-1}(\partial \mathcal{D}) = \frac{1}{b_{n-1}} \int_{\xi \in S_{n-1}} V_{n-1}(p_\xi(\mathcal{D})) \, dS. \tag{6.36}$$

In this expression S_{n-1} is the unit $(n-1)$-sphere, b_n the volume of the unit n-ball with $s_{n-1} = n \, b_n$, ξ is a unit vector, and p_ξ indicates the projection in the direction of ξ onto a hyperplane orthogonal to ξ. The second geometrical result applies to any compact q-dimensional domain $\mathcal{D} \in \mathbf{R}^{p+q}$, with $p \geq 2$, and states that its area $\mathcal{A}_q(\mathcal{D})$ obeys the equality

$$\mathcal{A}_q(\mathcal{D}) = \frac{s_p}{s_{p-1} s_{p+q}} \int_{\xi \in S_{p+q-1}} \mathcal{A}_{n-2}(p_\xi(\mathcal{D})) d\xi, \tag{6.37}$$

where ξ and p_ξ have the same significance as in the Cauchy formula (6.36). The Cauchy formula is then used to obtain the lower bound of (6.35) and the upper bound is a consequence of (6.37). If we now introduce into (6.35) the relation (6.30) between the mass M of the shell and the total mean curvature Q we find, since $n = D - 1$, that

$$\frac{s_{D-1}}{2 s_{D-3}} \Omega_{D-3} \leq \frac{16 \, G_D}{D - 2} M \leq \frac{s_{D-2}}{s_{D-3}} \omega_{D-3}. \tag{6.38}$$

The formulae

$$\frac{s_n}{s_{n-2}} = \frac{2\pi}{n-1} \quad \text{and} \quad \frac{s_{n-1}}{s_{n-2}} = \sqrt{\pi} \frac{\Gamma\left(\frac{n-1}{2}\right)}{\Gamma\left(\frac{n}{2}\right)} \omega_{D-3}, \tag{6.39}$$

can be used to rewrite (6.38) as

$$\frac{\pi}{D-2} \Omega_{D-3} \leq \frac{16 \, G_D}{D-2} M \leq \sqrt{\pi} \frac{\Gamma\left(\frac{D-2}{2}\right)}{\Gamma\left(\frac{D-1}{2}\right)} \omega_{D-3}. \tag{6.40}$$

The inequalities (6.40) specialize to the inequalities (5.41) in the particular case of $D = 4$. Since we have the dimensional relation $G_D M \sim \Omega_{D-3} \sim \omega_{D-3} \sim (\text{length})^{D-3}$, what was referred to as a hoop in general relativity becomes a $(D-3)$-dimensional closed strip in a space–time with D dimensions. In the brane-world models only one of the $D-3$ dimensions of the strip belongs to the brane; the $D-4$ remaining dimensions are in the bulk.

6.3 Light-like boost of higher dimensional black holes

Working with the Riemann curvature tensor, which is gauge invariant and unambiguously represents the gravitational field, is a mathematically sound approach to describe a Lorentz boosted gravitational field. Such an approach was adopted to obtain the gravitational field of an isolated gravitating body boosted to the velocity of light and was illustrated with many examples (Barrabès and Hogan 2003b) and used to derive scattering properties of high-speed sources (see Chapter 5). Since this method however requires lengthy calculations in four dimensions, only the weak field approximation will be considered for space–times with arbitrarily large dimensions. The validity of this approximation will be checked against the case of the $D = 4$ Kerr black hole for which the exact solution is known.

To begin with consider a black hole with mass M and angular momentum parameter a rotating around the z-axis of the Cartesian coordinates (x, y, z). The Lorentz boost can be implemented in an arbitrary direction (Barrabès and Hogan, 2004a) but for simplicity we only consider here the two cases corresponding to a longitudinal boost in the z-direction or to a transverse boost in (say) the x-direction. In both cases the mass of the black hole is rescaled as $M = p\gamma^{-1}$ where γ is the Lorentz factor. With such a rescaling the energy p stays finite since $M \to 0$ when $\gamma \to \infty$. The angular momentum parameter is rescaled differently according to the direction of the boost. A justification of this can be given by boosting a gravitating body with multipole moments and is described by Barrabès and Hogan (2003b) in the 4-dimensional case. For a transverse boost (in the direction of the x-axis) the angular momentum parameter a is not rescaled. Then since $M \to 0$ the angular momentum $J = Ma$ vanishes in the limit $\gamma \to \infty$ and we obtain the boosted Kerr space–time with line-element

$$ds^2 = -du\, dv + dy^2 + dz^2 - 4p\, \delta(u) \ln[(y-a)^2 + z^2]\, du^2, \qquad (6.41)$$

with $u = t - x$, and $v = t + x$. It is clear from this that although $J \to 0$ the influence of rotation is still present in this line-element through the appearance of the parameter a. The line-element (6.41) describes the space–time model of the gravitational field of a plane-fronted impulsive gravitational wave whose history is the null hyperplane with equation $u = 0$. The null generator of this hypersurface corresponding to $y = a, z = 0$ is a line singularity which is a remnant of the Kerr black hole. In comparison with the metric of a boosted Schwarzschild black hole [see (6.42)] the rotation produces a shift in this line singularity in the y-direction. This effect, known for a long time, has been checked by considering the deflection of highly relativistic particles in the Kerr gravitational field (Barrabès and Hogan, 2003b). For a longitudinal light-like boost (in the direction of the z-axis) the angular momentum parameter is rescaled as $\hat{a} = a\gamma$, with $\hat{a}$ finite. Hence $a \to 0$ when $\gamma \to \infty$ and the boosted Kerr space–time has line-element in this case given by

$$ds^2 = -du\, dv + dy^2 + dz^2 - 4p\, \delta(u) \ln(x^2 + y^2)\, du^2, \qquad (6.42)$$

with $u = t - z$ and $v = t + z$. This time the effect of rotation completely disappears and the boosted line-element is the Aichelburg–Sexl (1971) line-element, which is the line-element of the boosted Schwarzschild space–time. This line-element again describes the space–time model of the gravitational field of an impulsive plane gravitational wave with space–time history $u = 0$ and has a singularity on the null geodesic generator $y = z = 0$ of the null hypersurface $u = 0$.

We now compare these exact results with the expression for the boosted Kerr line-element which is obtained by imposing from the very beginning the weak field approximation (i.e. M and a small). We thus start with the following linearized form of the Kerr line-element

$$d\bar{s}^2 = \eta_{ab}\, d\bar{x}^a\, d\bar{x}^b + \frac{2M}{\bar{r}}(d\bar{t}^2 + d\bar{x}^2 + d\bar{y}^2 + d\bar{z}^2) + \frac{4Ma}{\bar{r}^3}(\bar{x}\, d\bar{y} - \bar{y}\, d\bar{x})\, d\bar{t}, \qquad (6.43)$$

where $\bar{x}^a = (\bar{x}, \bar{y}, \bar{z}, \bar{t})$, and $\bar{r}^2 = \bar{x}^2 + \bar{y}^2 + \bar{z}^2$. As before the bar refers to a quantity before making the Lorentz boost, and the same unbarred quantity corresponds to its boosted value after taking the limit $\gamma \to \infty$. The following two formulae will now be used and will also later be useful for dimensions $D > 4$:

$$\lim_{\gamma \to \infty} \frac{\gamma}{\sqrt{\gamma^2 u^2 + \rho^2}} = -2\ln|\rho|\delta(u) + \frac{1}{|u|}, \qquad (6.44)$$

and, for $m > 1$,

$$\lim_{\gamma \to \infty} \frac{\gamma}{(\gamma^2 u^2 + \rho^2)^{m/2}} = \sqrt{\pi} \frac{\Gamma(\frac{m-1}{2})}{\Gamma(\frac{m}{2})} \frac{\delta(u)}{\rho^{m-1}}. \qquad (6.45)$$

Here $\Gamma(x)$ is the gamma function satisfying the relations $\Gamma(x+1) = x\,\Gamma(x)$ for $x > 0$, $\Gamma(1) = \Gamma(0) = 1$ and $\Gamma(1/2) = \sqrt{\pi}$.

In the case of a longitudinal boost (i.e. for $\bar{t} = \gamma(t - vz)$, $\bar{z} = \gamma(z - vt)$, $\bar{x} = x$, $\bar{y} = y$) both M and a are rescaled. When $\gamma \to \infty$ we have $d\bar{z} \sim -\gamma du$ and $d\bar{t} \sim \gamma du$ and the line-element (6.43) becomes

$$ds^2 = \eta_{\alpha\beta} dx^\alpha dx^\beta + \frac{4p}{|u|} du^2 - 4p\ln(x^2 + y^2)\delta(u)\, du^2, \qquad (6.46)$$

with $u = t - z$. A coordinate transformation then brings this line-element into the form (6.42). As noted earlier the rescaling of a in this case wipes out the effect of rotation. For a transverse boost along the x-axis (i.e. for $\bar{t} = \gamma(t - vx)$, $\bar{x} = \gamma(x - vt)$, $\bar{y} = y$, $\bar{z} = z$) and in the limit $\gamma \to \infty$ we have $d\bar{x} \sim -\gamma du$ and $d\bar{t} \sim \gamma du$, with $u = t - x$. In this case only the mass M is rescaled and not a and the boosted line-element (6.43) is given by

$$ds^2 = -dudv + dy^2 + dz^2 + \frac{4p}{|u|} du^2 - 8p\,\delta(u)\left(\ln\rho - \frac{ay}{\rho^2}\right) du^2, \qquad (6.47)$$

with $\rho = \sqrt{y^2 + z^2}$, $u = t - x$, and $v = t + x$. It is easy to check that (6.47) is identical to the linearized form of (6.41) for small M and small a.

We now apply the weak field approximation to higher dimensional ($D > 4$) black holes, and consider rotating black holes, the non-rotating solution being simply obtained by equating to zero all the angular momentum parameters. The line-element is given by (6.10) with the additional relations (6.11)–(6.16). When the mass M of the black hole and the angular momentum parameters a_i are small an approximate form of (6.10) is the following:

$$d\bar{s}^2 = \eta_{ab}\, d\bar{x}^a d\bar{x}^b + 2\Phi \left[d\bar{t}^2 + \frac{d\bar{\sigma}^2}{D-3}\right] + 4\, d\bar{t}\, \sum_{i=1}^{l} A_i\,(\bar{x}^i\, d\bar{y}^i - \bar{y}^i\, d\bar{x}^i), \qquad (6.48)$$

where, as above, bars refer to quantities before the Lorentz boost. We have introduced the spatial line-element

$$d\bar{\sigma}^2 = \sum_{i=1}^{l} [(d\bar{x}^i)^2 + (d\bar{y}^i)^2] + \epsilon d\bar{z}^2, \qquad (6.49)$$

with $\epsilon = 0$ and $l = (D-1)/2$ when D is odd and with $\epsilon = 1$ and $l = (D-2)/2$ when D is even. The quantities Φ and A_i which appear in (6.48) are given by

$$\Phi = \frac{16\pi G_D M}{(D-2)s_{D-2}\,\bar{r}^{D-3}} = \frac{r_H^{D-3}}{2\,\bar{r}^{D-3}}, \qquad (6.50)$$

and

$$A_i = \frac{16\pi G_D J_i}{4s_{D-2}\,\bar{r}^{D-1}}, \qquad (6.51)$$

where r_H is found in (6.7) and

$$\bar{r}^2 = \sum_{i=1}^{l}[(\bar{x}^i)^2 + (\bar{y}^i)^2] + \epsilon \bar{z}^2. \qquad (6.52)$$

We recall that M is the mass of the black hole, $J_i = 2Ma_i/(D-2)$ is the angular momentum in the i-th biplane of rotation, and the area s_n of the unit sphere S^n is given in (6.6).

6.3.1 Space–times with odd dimension

For space–times with an odd number of dimensions we write the coordinates as $\bar{x}^a = (\bar{x}^1, \bar{y}^1, \ldots, \bar{x}^l, \bar{y}^l, \bar{t})$ with $l = (D-1)/2$. The direction of the boost is transversal as it necessarily lies within one of the biplanes of rotation. Therefore none of the angular momentum parameters is rescaled, a_i remains finite for $i = 1, 2, \ldots, l$ when the Lorentz factor $\gamma \to \infty$, and for the mass we have $M = p\gamma^{-1}$. If say the $\bar{x}^j$-axis is the direction of the Lorentz boost then

$$\bar{x}^j = \gamma(x^j - vt), \ \bar{t} = \gamma(t - vx^j), \ \bar{y}^j = y^j, \ \bar{x}^i = x^i, \ \bar{y}^i = y^i, \qquad (6.53)$$

for $i \neq j$. We define the retarded time coordinate $u = t - x^j$ in this case. When $\gamma \gg 1$ we have $\bar{x}^j \sim -\gamma u$ and $\bar{t} \sim \gamma u$. Then an approximate form of the boosted line-element is

$$ds^2 \sim \eta_{ab}\,dx^a dx^b + 2\left(\frac{D-2}{D-3}\right)\Phi\gamma^2 du^2 + 4A_j\gamma^2(-udy^j + y^j du)^2$$
$$+ \frac{2\Phi}{D-3}\left[(dy^j)^2 + \sum_{i \neq j}((dx^i)^2 + (dy^i)^2)\right] + 4\sum_{i \neq j}A_i\gamma(x^i dy^i - y^i dx^i)du. \qquad (6.54)$$

In the expressions (6.50) and (6.51) for Φ and A_i we have, since $\epsilon = 0$,

$$\bar{r}^2 \sim \gamma^2 u^2 + \rho^2, \qquad \rho^2 \equiv (y^j)^2 + \sum_{i \neq j}[(x^i)^2 + (y^i)^2]. \qquad (6.55)$$

The rescaling of M and the invariance of the a_i's imply that the last two terms of (6.54) become negligible in the light-like limit. Using (6.45) with $D > 4$ leads to

$$\lim_{\gamma \to \infty} 2\Phi\gamma^2 = \left(\frac{D-3}{D-2}\right)\frac{4\pi G_D\,p\,\Gamma(\frac{D-4}{2})}{\pi^{\frac{D-2}{2}}\rho^{D-4}}\delta(u), \qquad (6.56)$$

and

$$\lim_{\gamma \to \infty} 4 A_j \gamma^2 = \frac{8\pi G_D \, p \, a_j \, \Gamma(\frac{D-2}{2})}{\pi^{\frac{D-2}{2}} \rho^{D-2}} \delta(u). \tag{6.57}$$

Then the line-element (6.54) is given exactly, in the light-like limit, by

$$ds^2 = \eta_{ab} \, dx^a \, dx^b + \frac{4\pi G_D \, p \, \Gamma(\frac{D-4}{2})}{\pi^{\frac{D-2}{2}} \rho^{D-4}} \delta(u) \left[1 + \frac{D-4}{\rho^2} a_j y^j \right] du^2, \tag{6.58}$$

where the property $u \, \delta(u) = 0$ of the Dirac delta function has been used. It can be checked that the line-element (6.58) coincides with the line-element

$$ds^2 = \eta_{ab} \, dx^a \, dx^b + \frac{4\pi G_D \, p \, \Gamma(\frac{D-4}{2}) \delta(u) \, du^2}{\pi^{\frac{D-2}{2}} \left[(y^j - a_j)^2 + \sum_{i \neq j} [(x^i)^2 + (y^i)^2] \right]^{\frac{D-4}{2}}}, \tag{6.59}$$

when this line-element is linearized with respect to a^j. This latter expression shows that the effect of the rotation is to produce a shift along the y^j-axis perpendicular to the x^j-axis in the j-th 2-plane of rotation, an effect which already appeared when $D = 4$ [see (6.41)].

6.3.2 Space–times with even dimension

When the number of dimensions D is even the coordinates are given by the set $\bar{x}^a = (\bar{x}^1, \bar{y}^1, \ldots, \bar{x}^l, \bar{y}^l, \bar{z}, \bar{t},)$ with $l = (D-2)/2$. The direction of the boost can either be the z-axis or it can lie within one of the biplanes of rotation. In the first case the boost is of the longitudinal type and both M and the a_i's are rescaled according to $M = m\gamma^{-1}$ and $a_i = \hat{a}_i \gamma^{-1}$. In the second case it is transversal and only M is rescaled as $M = p\gamma^{-1}$. From (6.49) and (6.52) we have

$$d\bar{\sigma}^2 = \sum_{i=1}^{l} [(d\bar{x}^i)^2 + (d\bar{y}^i)^2] + d\bar{z}^2, \quad \bar{r}^2 = \sum_{i=1}^{l} [(\bar{x}^i)^2 + (\bar{y}^i)^2] + \bar{z}^2. \tag{6.60}$$

We consider first a transversal boost along the $\bar{x}^j$-axis. The notations used above when D is odd still apply, in particular for the relations (6.54) and (6.58), but with ρ now given by

$$\rho^2 \equiv (y^j)^2 + \sum_{i \neq j} [(x^i)^2 + (y^i)^2] + z^2. \tag{6.61}$$

We find that the boosted line-element reads

$$ds^2 = \eta_{ab} \, dx^a \, dx^b + \frac{4\pi G_D p \, \Gamma(\frac{D-4}{2})}{\pi^{\frac{D-2}{2}} \rho^{D-4}} \delta(u) \left[1 + \frac{D-4}{\rho^2} a_j y^j \right] du^2. \tag{6.62}$$

In parallel with the case of odd dimensions, (6.62) is identical with the linearized form of the line-element

$$ds^2 = \eta_{ab} \, dx^a \, dx^b + \frac{4\pi G_D p \, \Gamma(\frac{D-4}{2}) \delta(u) \, du^2}{\pi^{\frac{D-2}{2}} \left[(y^j - a_j)^2 + \sum_{i \neq j} [(x^i)^2 + (y^i)^2] + z^2 \right]^{\frac{D-4}{2}}}. \tag{6.63}$$

On the other hand, for a boost in the direction of the $\bar{z}$-axis,
$$\bar{z} = \gamma(z - vt), \quad \bar{t} = \gamma(t - vz), \quad \bar{x}^i = x^i, \quad \bar{y}^i = y^i, \tag{6.64}$$
for $i = 1, 2, \ldots, l$ and thus when $\gamma \gg 1$ we have $\bar{z} \sim -\gamma u$ and $\bar{t} \sim \gamma u$, with $u = t - z$. Hence the boosted line-element is given approximately by
$$ds^2 \sim \eta_{ab}\, dx^a\, dx^b + 2\left(\frac{D-2}{D-3}\right)\Phi\gamma^2 du^2 + \frac{2\Phi}{D-3}\left[\sum_{i=1}^{l}((dx^i)^2 + (dy^i)^2)\right]$$
$$+ 4\sum_{i=1}^{l} A_i \gamma (x^i\, dy^i - y^i\, dx^i)\, du. \tag{6.65}$$
In the expressions for Φ and the A_i's we have
$$\bar{r}^2 \sim \gamma^2 u^2 + \rho^2 \quad \text{and} \quad \rho^2 \equiv \sum_{i=1}^{l}[(x^i)^2 + (y^i)^2]. \tag{6.66}$$
Since the boost is in the z-direction and is longitudinal all the angular momentum parameters scale according to $a_i = \hat{a}_i \gamma^{-1}$ with $\hat{a}_i$ finite in the limit $\gamma \to \infty$. The expression (6.56) still applies but we have $\lim_{\gamma \to \infty} \Phi = \lim_{\gamma \to \infty} A_i \gamma = 0$, and the last two terms of (6.65) disappear. The boosted line-element is thus
$$ds^2 = \eta_{ab}\, dx^a\, dx^b + \frac{4\pi G_D\, p\, \Gamma(\frac{D-4}{2})}{\pi^{\frac{D-2}{2}}\, \rho^{D-4}}\, \delta(u)\, du^2, \tag{6.67}$$
with ρ given by (6.66). The effect of rotation has disappeared in the longitudinal boost as was the case when $D = 4$. Equation (6.67) represents the ultra-relativistic limit of the boosted Tangherlini black hole solution (6.4) for $D > 4$.

Appendix A
Notation

We give a brief summary here of the notation and sign conventions used in this book. Throughout we use units in which the speed of light in a vacuum is $c = 1$ and the gravitational constant is $G = 1$. We use a metric of signature $+2$. Latin indices take values 1, 2, 3, 4 and Greek indices take values 1, 2, 3. Latin letters from the second half of the alphabet will generally denote coordinate components of tensor fields on space–time. In a local coordinate system x^i the components of the metric tensor are (g_{ij}) and the line-element of space–time is

$$ds^2 = g_{ij}\, dx^i\, dx^j. \tag{A.1}$$

Partial derivatives are indicated by a comma (for example $f_{,i} = \partial f/\partial x^i$) and covariant derivatives by a semicolon (for example $v_{i;j} = v_{i,j} - \Gamma^k_{ij} v_k$ where Γ^k_{ij} are the components of the Riemannian connection). Latin letters from the first half of the alphabet will generally denote tetrad components starting with the introduction of a set of basis 1-forms ϑ^a. In terms of these we can write

$$ds^2 = g_{ab}\, \vartheta^a\, \vartheta^b, \tag{A.2}$$

where g_{ab} are now the components of the metric tensor on the tetrad basis defined via the 1-forms. We will always normalize the basis so that g_{ab} are constants. Denoting exterior differentiation by d, the components $\omega_{ab} = -\omega_{ba}$ of the Riemannian connection 1-form are given by the first Cartan structure equation:

$$d\vartheta^a = -\omega^a{}_b \wedge \vartheta^b. \tag{A.3}$$

Here tetrad indices $a, b, c, \ldots$ are raised with g^{ab}, where $g^{ab} g_{bc} = \delta^a_c$ and so g^{ab} are the components of the inverse of the matrix with entries g_{ab}. Thus $\omega^a{}_b = g^{ac} \omega_{cb}$. Tetrad indices are lowered with g_{ab}. The curvature 2-form $\Omega_{ab} = -\Omega_{ba}$ is obtained from the second Cartan structure equation:

$$\Omega_{ab} = d\omega_{ab} + \omega_{ac} \wedge \omega^c{}_b. \tag{A.4}$$

From this we obtain the tetrad components of the Riemann curvature tensor R_{abcd} from

$$\Omega_{ab} = \frac{1}{2} R_{abcd}\, \vartheta^c \wedge \vartheta^d. \tag{A.5}$$

The Ricci tensor has components $R_{ab} = g^{cd} R_{acbd}$ and the Ricci scalar is $R = g^{ab} R_{ab}$. From these the components C_{abcd} of the Weyl conformal curvature tensor are

$$C_{abcd} = R_{abcd} + \frac{1}{2}\left(g_{ad} R_{bc} + g_{bc} R_{ad} - g_{ac} R_{bd} - g_{bd} R_{ac}\right) + \frac{1}{6} R \left(g_{ac} g_{bd} - g_{ad} g_{bc}\right). \tag{A.6}$$

We often specialize (A.2) to

$$ds^2 = (\vartheta^1)^2 + (\vartheta^2)^2 - 2\,\vartheta^3\,\vartheta^4. \tag{A.7}$$

In this case the basis of 1-forms defines a 'half-null' tetrad meaning that the tetrad consists of two space-like vectors and two null vectors. On such a tetrad the Newman–Penrose components of the Weyl conformal curvature tensor are given by

$$\Psi_0 = R_{1313} - \frac{1}{2} R_{33} + i R_{1323}, \tag{A.8}$$

$$\Psi_1 = \frac{1}{\sqrt{2}} \left(R_{3431} + i R_{3432}\right) - \frac{1}{2\sqrt{2}} \left(R_{31} + i R_{32}\right), \tag{A.9}$$

$$\Psi_2 = \frac{1}{2} \left(R_{3434} + i R_{3412} - R_{34} + \frac{1}{6} R\right), \tag{A.10}$$

$$\Psi_3 = \frac{1}{\sqrt{2}} \left(R_{3414} - i R_{3424} + \frac{1}{2} R_{41} + \frac{1}{2} i R_{42}\right), \tag{A.11}$$

$$\Psi_4 = R_{1414} - \frac{1}{2} R_{44} - i R_{1424}, \tag{A.12}$$

with $R = R_{11} + R_{22} - 2 R_{34}$.

Appendix B
Transport law for k along r = 0

In place of (3.15) let us write

$$P k^i = -\xi \lambda^i_{(1)} - \eta \lambda^i_{(2)} - \left(1 - \frac{1}{4}(\xi^2 + \eta^2)\right) \lambda^i_{(3)} + \left(1 + \frac{1}{4}(\xi^2 + \eta^2)\right) \lambda^i_{(4)}, \quad \text{(B.1)}$$

with $\{\lambda^i_{(b)}(u)\}^4_{b=1}$ an orthonormal tetrad defined along the world line $r = 0$ with $\lambda^i_{(4)} = v^i$ and the orthonormal triad $\{\lambda^i_{(\alpha)}\}^3_{\alpha=1}$ satisfying $\lambda^i_{(\alpha)} v_i = 0$ for $\alpha = 1, 2, 3$. Thus on account of the second of (3.14) the function P in (B.1) is given by

$$P = 1 + \frac{1}{4}(\xi^2 + \eta^2). \quad \text{(B.2)}$$

We assume that $\{\lambda^i_{(\alpha)}\}^3_{\alpha=1}$ is transported along $r = 0$ according to

$$\frac{\partial \lambda^i_{(\alpha)}}{\partial u} = (v^i a^j - v^j a^i) \lambda_{(\alpha)j} + \omega^{ij} \lambda_{(\alpha)j}, \quad \text{(B.3)}$$

with $\omega_{ij}(u) = -\omega_{ji}(u)$ and $\omega_{ij}(u) v^j(u) = 0$. If $\omega_{ij} = 0$ then this transport law becomes Fermi–Walker transport while the final term in (B.3) supplies a spatial rotation to the triad $\{\lambda^i_{(\alpha)}\}^3_{\alpha=1}$ (Misner et al. 1973, p. 174). It follows from (B.3) that

$$\omega_{(\alpha\beta)} \equiv \omega_{ij} \lambda^i_{(\alpha)} \lambda^j_{(\beta)} = \lambda_{(\alpha)i} \dot\lambda^i_{(\beta)} = -\omega_{(\beta\alpha)}, \quad \text{(B.4)}$$

with the dot, as always, denoting differentiation with respect to u. From (B.1) and (B.2) we see that (B.3) implies the transport law

$$\frac{\partial k^i}{\partial u} = (v^i a^j - v^j a^i) k_j + \omega^{ij} k_j, \quad \text{(B.5)}$$

for k^i along $r = 0$. This transport law preserves (3.14) along $r = 0$ just as the transport law (3.17) did. Using (B.5) in conjunction with the transformation (3.13) we have

$$dx^i = \{(1 + r a_j k^j) v^i + r a^i + r \omega^{ij} k_j\} du + k^i dr + r \left(\frac{\partial k^i}{\partial \xi} d\xi + \frac{\partial k^i}{\partial \eta} d\eta\right). \quad \text{(B.6)}$$

In utilizing this to evaluate the Minkowskian space–time line-element in coordinates ξ, η, r, u a useful identity is found to be

$$P^2 \left|(a_i + \omega_{ij} k^j)\frac{\partial k^i}{\partial \xi} - i(a_i + \omega_{ij} k^j)\frac{\partial k^i}{\partial \eta}\right|^2 = (a_i + \omega_{ij} k^j)(a^i + \omega^{il} k_l) - (a_i k^i)^2. \quad \text{(B.7)}$$

Using this with (B.6) we obtain

$$\eta_{ij}\,dx^i\,dx^j = r^2 P^{-2}\left\{\left(d\xi + P^2\left(a_i + \omega_{ij}\,k^j\right)\frac{\partial k^i}{\partial \xi}du\right)^2\right.$$

$$\left. + \left(d\eta - P^2\left(a_i + \omega_{ij}\,k^j\right)\frac{\partial k^i}{\partial \eta}du\right)^2\right\}$$

$$- 2\,du\,dr - (1 - 2H\,r)\,du^2, \tag{B.8}$$

with $H = -a_j\,k^j$ and P given by (B.2). With k^i given by (B.1) we have

$$H = -a_i\,k^i = -P^{-1}\left\{\xi\,a_{(1)} + \eta\,a_{(2)} + \left(1 - \frac{1}{4}(\xi^2 + \eta^2)\right)a_{(3)}\right\}, \tag{B.9}$$

where $a_{(\alpha)}(u) = a_i(u)\,\lambda^i_{(\alpha)}(u)$ for $\alpha = 1, 2, 3$. If we now define the function

$$q(\xi, \eta, u) = -\eta\left(1 - \frac{1}{4}\xi^2 + \frac{1}{12}\eta^2\right)a_{(1)} - \xi\left(1 + \frac{1}{12}\xi^2 - \frac{1}{4}\eta^2\right)a_{(2)}$$

$$+ \xi\eta\,a_{(3)} + \frac{1}{2}(\xi^2 - \eta^2)\omega_{(12)} - \eta\left(1 + \frac{1}{4}\xi^2 - \frac{1}{12}\eta^2\right)\omega_{(13)}$$

$$- \xi\left(1 - \frac{1}{12}\xi^2 + \frac{1}{4}\eta^2\right)\omega_{(23)}, \tag{B.10}$$

with $\omega_{(\alpha\beta)}(u)$ given by (B.4), then (B.8) can be rewritten in the Robinson–Trautman (1960, 1962) form

$$\eta_{ij}\,dx^i\,dx^j = r^2 P^{-2}\left\{\left(d\xi - \frac{\partial q}{\partial \eta}du\right)^2 + \left(d\eta - \frac{\partial q}{\partial \xi}du\right)^2\right\} - 2\,du\,dr - (1 - 2H\,r)\,du^2. \tag{B.11}$$

This form of the Minkowskian space–time line-element was first published by Molenda (1984). In some situations it is more useful than the form (3.21) [see, for example Hogan and Ellis (1989)].

Appendix C
Some useful scalar products

For Chapter 3 the following list of scalar products involving the vector field U^i defined in (3.114) are useful in calculating the line-element (3.119):

$$\frac{\partial k_i}{\partial \xi}\frac{\partial U^i}{\partial \xi} = 0, \tag{C.1}$$

$$\frac{\partial k_i}{\partial \eta}\frac{\partial U^i}{\partial \xi} = -P_0^{-2}F, \tag{C.2}$$

$$v_i \frac{\partial U^i}{\partial \xi} = 0, \tag{C.3}$$

$$k_i \frac{\partial U^i}{\partial \xi} = F_\eta, \tag{C.4}$$

$$\frac{\partial k_i}{\partial \eta}\frac{\partial U^i}{\partial \eta} = 0, \tag{C.5}$$

$$\frac{\partial k_i}{\partial \xi}\frac{\partial U^i}{\partial \eta} = P_0^{-2}F, \tag{C.6}$$

$$v_i \frac{\partial U^i}{\partial \eta} = 0, \tag{C.7}$$

$$k_i \frac{\partial U^i}{\partial \eta} = -F_\xi, \tag{C.8}$$

$$\frac{\partial U_i}{\partial u}\frac{\partial U^i}{\partial u} = -F^2 P_0^2 \left(\left(\frac{\partial h_0}{\partial \xi}\right)^2 + \left(\frac{\partial h_0}{\partial \eta}\right)^2\right), \tag{C.9}$$

$$v_i \frac{\partial U^i}{\partial u} = -P_0^2 \left(\frac{\partial h_0}{\partial \xi}F_\eta - \frac{\partial h_0}{\partial \eta}F_\xi\right), \tag{C.10}$$

$$\frac{\partial U_i}{\partial u}\frac{\partial U^i}{\partial \xi} = -P_0^2 \left(\frac{\partial h_0}{\partial \xi}F_\eta - \frac{\partial h_0}{\partial \eta}F_\xi\right)F_\eta - \frac{\partial h_0}{\partial \xi}F^2, \tag{C.11}$$

$$\frac{\partial U_i}{\partial u}\frac{\partial U^i}{\partial \eta} = P_0^2 \left(\frac{\partial h_0}{\partial \xi}F_\eta - \frac{\partial h_0}{\partial \eta}F_\xi\right)F_\xi - \frac{\partial h_0}{\partial \eta}F^2, \tag{C.12}$$

$$k_i \frac{\partial U^i}{\partial u} = 0, \tag{C.13}$$

$$\frac{\partial k_i}{\partial \xi}\frac{\partial U^i}{\partial u} = \frac{\partial h_0}{\partial \eta}F, \tag{C.14}$$

Some useful scalar products

$$\frac{\partial k_i}{\partial \eta}\frac{\partial U^i}{\partial u} = -\frac{\partial h_0}{\partial \xi}F,\qquad\text{(C.15)}$$

$$\frac{\partial U_i}{\partial \xi}\frac{\partial U^i}{\partial \xi} = -P_0^{-2}F^2 - F_\eta^2,\qquad\text{(C.16)}$$

$$\frac{\partial U_i}{\partial \eta}\frac{\partial U^i}{\partial \eta} = -P_0^{-2}F^2 - F_\xi^2,\qquad\text{(C.17)}$$

$$\frac{\partial U_i}{\partial \xi}\frac{\partial U^i}{\partial \eta} = F_\xi F_\eta.\qquad\text{(C.18)}$$

References

Aichelburg, P. C. and Sexl, R. U. (1971). *Gen. Rel. Grav.*, **2**, 303–312.
Antoniadis, L., Arkani-Hamed, N., Dimopoulos, S., and Dvali, G. (1998). *Phys. Lett. B*, **436**, 257–263.
Asada, H., Futamase, T., and Hogan, P. A. (2010). *Equations of Motion in General Relativity*. Oxford University Press, Oxford.
Balasin, H. and Nachbagauer, H. (1996). *Class. Quantum Grav.*, **13**, 731–737.
Barrabès, C. and Frolov, V. P. (1996). *Phys. Rev. D*, **53**, 3215–3223.
Barrabès, C., Frolov, V. P., and Hogan, P. A. (2005). *Class. Quantum Grav.*, **22**, 2085–2100.
Barrabès, C., Frolov, V. P., and Lesigne, E. (2004). *Phys. Rev. D*, **69** (101501).
Barrabès, C., Frolov, V. P., and Parentani, R. (1999). *Phys. Rev. D*, **59** (124010).
Barrabès, C., Gramain, A., Lesigne, E., and Letelier, P. S. (1992). *Class. Quantum Grav.*, **9**, L105–L110.
Barrabès, C. and Hogan, P. A. (2001). *Phys. Rev. D*, **64** (044022).
Barrabès, C. and Hogan, P. A. (2003a). *Phys. Rev. D*, **67** (084028).
Barrabès, C. and Hogan, P. A. (2003b). *Singular Null Hypersurfaces in General Relativity*. World Scientific, Singapore.
Barrabès, C. and Hogan, P. A. (2004a). *Phys. Rev. D*, **70** (107502).
Barrabès, C. and Hogan, P. A. (2004b). *Class. Quantum Grav.*, **21**, 405–416.
Barrabès, C. and Hogan, P. A. (2007). *Phys. Rev. D*, **75** (124012).
Barrabès, C. and Hogan, P. A. (2011). *Prog. Theor. Phys.*, **126**, 1157–1165.
Barrabès, C. and Israel, W. (1991). *Phys. Rev. D*, **58**, 1129–1142.
Barrabès, C., Israel, W., and Poisson, E. (1990). *Class. Quantum Grav.*, **7**, L273–L278.
Bell, P. and Szekeres, P. (1974). *Gen. Rel. Grav.*, **5**, 275–286.
Bertotti, B. (1959). *Phys. Rev.*, **116**, 1331–1333.
Birrel, N. D. and Davies, P. C. W. (1982). *Quantum Fields in Curved Space*. Cambridge University Press, Cambridge.
Bolgar, Florian (2012). Report on Internship in University College Dublin. Technical report, École Normale Superieure Paris.
Bondi, H., van der Burgh, M. G. J., and Metzner, A. W. K. (1962). *Proc. R. Soc. A*, **269**, 21–52.
Boyer, R. H. and Lindquist, R. W. (1967). *J. Math. Phys.*, **8**, 265–281.
Brout, R., Massar, S., Parentani, R., and Spindel, Ph. (1995). *Phys. Rev. D*, **52**, 4559–4568.
Burago, Y. D. and Zalgaller, V. A. (1988). *Geometric Inequalities*. Springer, Berlin.
Burnett, G. A. (1989). *J. Math. Phys.*, **30**, 90–96.
Cardoso, V., Berti, E., and Cavaglià, M. (2005). *Class. Quantum Grav.*, **22**, L61–L69.

Carroll, S. M. (2004). *Spacetime and Geometry: An Introduction to General Relativity.* Addison Wesley, San Francisco.
Chandrasekhar, S. (1983). *The Mathematical Theory of Black Holes.* Oxford University Press, Oxford.
Choquet-Bruhat, Y. (1969). *Commun. Math. Phys.*, **12**, 16–35.
Corinaldesi, E. and Papapetrou, A. (1951). *Proc. R. Soc. A*, **209**, 259–268.
d'Inverno, Ray (1992). *Introducing Einstein's Relativity.* Clarendon Press, Oxford.
Dixon, W. G. (1970a). *Proc. R. Soc. A*, **314**, 499–527.
Dixon, W. G. (1970b). *Proc. R. Soc. A*, **319**, 509–547.
Dixon, W. G. (1973). *Gen. Rel. Grav.*, **4**, 199–209.
Dixon, W. G. (2008). *Acta Physica Polonica B (Proceedings Supplement)*, **1**, 27–54.
Dray, T. and 't Hooft, G. (1985). *Comm. Math. Phys.*, **99**, 613–625.
Eardley, D. M. and Giddings, S. B. (2002). *Phys. Rev. D*, **66** (044011).
Ehlers, J. (1993). *Gen. Rel. Grav. (reprinted from Abh. Akad. Wiss. Lit. Mainz (1961))*, **25**, 1225–1266.
Ellis, G. F. R. (1971). *Relativistic Cosmology.* Gordon and Breach, London, pp. 1–60.
Ellis, G. F. R. and Bruni, M. (1989). *Phys. Rev. D*, **40**, 1804–1818.
Ellis, G. F. R. and Hogan, P. A. (1997). *Gen. Rel. Grav.*, **29**, 235–244.
Emparan, R. and Myers, R. C. (2003). *J. High Energy Phys.*, **9** (025).
Emparan, R. and Myers, R. C. (2008). *Living Rev. Relativity*, **11** (6).
Emparan, R. and Reall, H. S. (2002). *Phys. Rev. Lett.*, **88** (101101).
Frolov, V. P. (2003). *Phys. Rev. D*, **67** (084004).
Frolov, V. P. and Fursaev, D. (2005). *Phys. Rev. D*, **71** (104034).
Frolov, V. P., Israel, W., and Zelnikov, A. (2005). *Phys. Rev. D*, **72** (084031).
Frolov, V. P. and Kubiznak, D. (2007). *Phys. Rev. Lett.*, **98** (011101).
Frolov, V. P., Markov, M. A., and Mukhanov, V. F. (1990). *Phys. Rev. D*, **41**, 383–394.
Frolov, V. P. and Novikov, I. D. (1998). *Black Hole Physics: Basic Concepts and New Developments.* Kluwer Academic Publishers, Dordrecht.
Frolov, V. P. and Sojkovic, D. (2003). *Phys. Rev. D*, **68** (064011).
Frolov, V. P. and Zelnikov, A. (2011). *Introduction to Black Hole Physics.* Oxford University Press, Oxford.
Futamase, T. and Hogan, P. A. (1993). *J. Math. Phys.*, **34**, 154–169.
Futamase, T., Hogan, P. A., and Itoh, Y. (2008). *Phys. Rev. D*, **78** (104014).
Gibbons, G. W. (1972). *Comm. Math. Phys.*, **27**, 87–102.
Gibbons, G. W. (1997). *Class. Quantum Grav.*, **14**, 2905–2915.
Gralla, S. E. and Wald, R. M. (2008). *Class. Quantum Grav.*, **25** (205009).
Griffiths, J. B. (1991). *Colliding Plane Waves in General Relativity.* Clarendon Press, Oxford.
Hawking, S. W. (1966). *Astrophys. J.*, **145**, 544–554.
Hawking, S. W. (1967). *Proc. R. Soc. A*, **300**, 182–201.
Hawking, S. W. (1971). *Phys. Rev. Lett.*, **26**, 1344–1346.
Hawking, S. W. (1975). *Comm. Math. Phys.*, **43**, 199–220.
Hawking, S. W. (1990). *Phys. Lett. B*, **246**, 36–38.
Hawking, S. W. and Ellis, G. F. R. (1973). *The Large Scale Structure of Space–Time.* Cambridge University Press, Cambridge.

Hogan, P. A. (1988). *Astrophys. J.*, **333**, 64–67.
Hogan, P. A. (1994). *Phys. Rev. D*, **49**, 6521–6525.
Hogan, P. A. and Ellis, G. F. R. (1989). *Ann. Phys. N.Y.*, **195**, 293–323.
Hogan, P. A. and Ellis, G. F. R. (1997). *Class. Quantum Grav.*, **14**, A171–A188.
Hogan, P. A. and O'Farrell, S. (2011). *Gen. Rel. Grav.*, **43**, 1625–1638.
Hogan, P. A. and O'Shea, E. M. (2002a). *Phys. Rev. D*, **65** (124017).
Hogan, P. A. and O'Shea, E. M. (2002b). *Phys. Rev. D*, **66** (124016).
Hogan, P. A. and Trautman, A. (1987). *Gravitation and Geometry*. Bibliopolis, Naples, pp. 215–242.
Horowitz, G. T. (2012). *Black Holes in Higher Dimensions*. Cambridge University Press, Cambridge.
Isaacson, R. A. (1968a). *Phys. Rev.*, **166**, 1263–1271.
Isaacson, R. A. (1968b). *Phys. Rev.*, **166**, 1272–1280.
Israel, W. and Poisson, E. (1988). *Class. Quantum Grav.*, **5**, L201–L205.
Kaluza, T. (1921). *Sitzungsber. Preuss. Akad. Wiss. Berlin*, **K1**, 966–972.
Kasner, E. (1925). *Trans. American Math. Soc.*, **27**, 155–162.
Kerr, R. P. (1963). *Phys. Rev. Lett.*, **11**, 237–238.
Kerr, R. P. and Schild, A. (1965a). Atti del Convegno sulla Relatività Generale: Problemi dell Energia e Onde Gravitazionali (Anniversary Volume, Fourth Centenary of Galileo's Birth). Firenze, pp. 222–233. Comitato Nazionale per le Manifestazione Celebrative.
Kerr, R. P. and Schild, A. (1965b). Proceedings of Symposium in Applied Mathematics. Volume XVII, Providence, RI, pp. 199–209. American Mathematical Society.
Khan, K. A. and Penrose, R. (1971). *Nature*, **229**, 185–186.
Klein, O. (1926). *Zeit. f. Physik*, **37**, 895–906.
Krasiński, A. (2006). *Inhomogeneous Cosmological Models*. Cambridge University Press, Cambridge.
Lorentz, H. A. (1923). *The Principle of Relativity*. Methuen, London.
Lukács, B., Perjés, Z., Porter, J., and Sebestyén, A. (1984). *Gen. Rel. Grav.*, **16**, 691–701.
MacCallum, M. A. H. and Taub, A. H. (1973). *Commun. Math. Phys.*, **30**, 153–169.
Markov, M. A. (1984). *Ann. Phys. N.Y.*, **155**, 333–357.
Markov, M. A. and Mukhanov, V. F. (1985). *Nuovo Cimento B*, **86**, 97–102.
Mathisson, M. (1937). *Acta Physica Polonica*, **6**, 163–200.
Minkowski, H. (1903). *Math. Ann.*, **57**, 447–495.
Minkowski, H. (1909). *Phys. Zeitschr.*, **10**, 104–111.
Misner, C. W., Thorne, K. S., and Wheeler, J. A. (1973). *Gravitation*. Freeman, San Francisco.
Molenda, T. (1984). Contributed Papers: Proceedings of 10th. International Conference on General Relativity and Gravitation 1983. Reidel, Dordrecht, pp. 97–99.
Mukhanov, V. F. and Brandenberger, R. (1992). *Phys. Rev. Lett.*, **68**, 1969–1972.
Myers, R. C. and Perry, M. J. (1986). *Ann. Phys. N.Y.*, **172**, 304–347.
Nariai, H. (1999). *Gen. Rel. Grav. (reprinted from Reports of Tohoku University (1951))*, **31**, 963–971.
Newman, E. T. and Penrose, R. (1962). *J. Math. Phys.*, **3**, 566–578.

Newman, E. T. and Unti, T. W. J. (1962). *J. Math. Phys.*, **3**, 891–901.
Newman, E. T. and Unti, T. W. J. (1963). *J. Math. Phys.*, **4**, 1467–1469.
Olmsted, J. M. H. (1959). *Real Variables*. Appleton-Century-Crofts, New York.
O'Shea, E. M. (2004a). *Phys. Rev. D*, **69** (064038).
O'Shea, E. M. (2004b). *Phys. Rev. D*, **70** (024001).
Ozsváth, I., Robinson, I., and Rózga, K. (1985). *J. Math. Phys.*, **26**, 1755–1761.
Papapetrou, A. (1951). *Proc. R. Soc. A*, **209**, 248–258.
Pelath, M. A., Tod, K. P., and Wald, R. (1998). *Class. Quantum Grav.*, **15**, 3917–3934.
Penrose, R. (1965). *Phys. Rev. Lett.*, **10**, 57–59.
Penrose, R. (1968). *Batelles Rencontres*. W. A. Benjamin, New York, pp. 121–235.
Penrose, R. (1972). *General Relativity: Papers in Honour of J. L. Synge*. Clarendon Press, Oxford, pp. 101–115.
Penrose, R. (1973). *Ann. N. Y. Acad. Sci.*, **224**, 125–134.
Penrose, R. and Rindler, W. (1984). *Spinors and Space–Time*, Vol. 1. Cambridge University Press, Cambridge.
Poisson, E. (2004). *A Relativist's Toolkit: The Mathematics of Black Hole Mechanics*. Cambridge University Press, Cambridge.
Poisson, E. and Israel, W. (1989). *Phys. Rev. Lett.*, **63**, 1663–1666.
Poisson, E. and Israel, W. (1990). *Phys. Rev. D*, **41**, 1796–1809.
Poisson, E., Pound, A., and Vega, I. (2011). *Living Rev. Relativity*, **14** (7).
Polchinski, J. (1989). *Nucl. Phys. B*, **325**, 619–630.
Pound, A. (2010). *Phys. Rev. D*, **81** (024023).
Price, R. H. (1972a). *Phys. Rev. D*, **5**, 2419–2438.
Price, R. H. (1972b). *Phys. Rev. D*, **5**, 2439–2454.
Randall, L. and Sundrum, R. (1999). *Phys. Rev. Lett.*, **83**, 3370–3373.
Redmount, I. H. (1985). *Prog. Theor. Phys.*, **73**, 1401–1426.
Rindler, W. and Trautman, A. (1987). *Gravitation and Geometry*. Bibliopolis, Naples.
Robinson, I. (1959). *Bull. Acad. Polon.*, **7**, 351–352.
Robinson, I. and Robinson, J. R. (1972). *General Relativity: Papers in Honour of J. L. Synge*. Clarendon Press, Oxford, pp. 151–166.
Robinson, I. and Trautman, A. (1960). *Phys. Rev. Lett.*, **4**, 431–432.
Robinson, I. and Trautman, A. (1962). *Proc. R. Soc. A*, **265**, 463–473.
Robinson, I. and Trautman, A. (1983). *J. Math. Phys.*, **24**, 1425–1429.
Rosen, N. (1937). *Phys. Z. Sowjet*, **12**, 366–372.
Rychkov, V. S. (2004). *Phys. Rev. D*, **70** (044003).
Sachs, R. (1962). *Proc. R. Soc. A*, **270**, 103–126.
Sauer, T. and Trautman, A. (2008). *Acta Physica Polonica B (Proceedings Supplement)*, **1**, 7–26.
Stephani, H., Kramer, D., MacCallum, M. A. H., Hoenselaers, C., and Herlt, E. (2003). *Exact Solutions of Einstein's Equations*, 2nd edn. Cambridge University Press, Cambridge.
Synge, J. L. (1964). *Relativity: The General Theory*. North Holland, Amsterdam.
Synge, J. L. (1965). *Relativity: The Special Theory*. North Holland, Amsterdam.
Synge, J. L. (1970). *Annali di Matematica Pura ed Applicata*, **84**, 33–60.
Szekeres, P. (1970). *Nature*, **228**, 1183–1184.

Szekeres, P. (1972). *J. Math. Phys.*, **13**, 286–294.
't Hooft, G. (1985). *Nucl. Phys. B*, **256**, 727–745.
't Hooft, G. (1996). *Int. J. Mod. Phys. A*, **11**, 4623–4688.
Tangherlini, F. R. (1963). *Nuovo Cimento*, **27**, 636–651.
Tod, K. P. (1992). *Class. Quantum Grav.*, **9**, 1581–1591.
Tran, H. V. (1988). *The Geometry of Plane Waves in Spaces of Constant Curvature*. PhD thesis, The University of Texas at Dallas.
Trautman, A. (1962). *Recent Developments in General Relativity*. PWN, Warsaw, pp. 459–463.
Trautman, A., Pirani, F. A. E., and Bondi, H. (1965). *Lectures on General Relativity*. Prentice-Hall, New Jersey.
Weinberg, S. (1972). *Gravitation and Cosmology*. John Wiley, New York.
York, J. W. (1983). *Phys. Rev. D*, **28**, 2929–2945.
Yoshino, H. and Nambu, Y. (2002). *Phys. Rev. D*, **66** (065004).
Yoshino, H. and Rychkov, V. S. (2005). *Phys. Rev. D*, **71** (044028).
Zel'dovich, Ya. B. and Novikov, I. D. (1978). *Relativistic Astrophysics, Volume 1: Stars and Relativity*. University of Chicago Press, Chicago.

Index

$\dot{E}$ equation, 63
$\dot{H}$ equation, 63
1-forms, 31, 34, 46, 70, 98, 131
2-form, 92
3-velocity, 7
4-acceleration, 30, 61
4-velocity, 6, 60
5-dimensional flat manifold, 57

A
advanced time, 105, 110
affine parameter, 5, 28
Aichelburg–Sexl metric, 95
angular momentum, 92, 103, 127
angular momentum parameter, 82, 125
angular velocity of Kerr black hole, 84
apparent horizon, 86, 89, 90, 123
area of apparent horizon, 88, 91
area of event horizon, 112
area of unit sphere, 120, 123, 127
area theorem, 88
asymptotic spectrum of Hawking radiation, 115, 117
asymptotically flat, 79

B
background Minkowskian space–time, 30
background Ricci tensor, 26, 27, 47
background space–time, 33, 45, 63
basis 1-forms, 17, 22, 24, 43, 131
Bel–Robinson tensor, 77
Bell–Szekeres solution, 19, 20
Bertotti–Robinson solution, 20
Bianchi identities, 61, 62, 65
blueshift, 104
Bogoliubov coefficients, 116
boosted Kerr space–time, 125
Boyer–Lindquist coordinates, 82, 121

C
Cartan structure equation, 131
Carter constant, 85
Carter–Penrose diagram, 81–84, 111
Cartesian coordinates, 1
Cartesian product, 20
Cauchy formula, 123
Cauchy horizon, 83, 86, 104
causal structure, 86, 120
collapse of a circular loop, 91
collapsing null shell, 111, 112
colliding null shells, 108, 109
complex shear, 27, 34, 102
conical singularity, 54
constant curvature, 20, 21, 58
convex closed surface, 88
convex domain, 88, 123
convex shells, 87
convex thin shell, 87, 122
coordinate conditions, 10
coordinate singularity, 80, 83
cosmic background radiation, 56, 78
cosmic censorship, 87, 88, 103
cosmological constant, 21, 23, 56, 59
cosmological perturbation, 60
creation of de Sitter universes, 79
curvature 2-form, 131
curvature invariant, 80, 83, 105

D
D-dimensional black hole, 119
D-dimensional form of Gibbons–Penrose inequality, 123
D-dimensional rotating black hole, 120
dark energy, 23
de Sitter horizon, 108
de Sitter phase bubble, 108, 110
de Sitter universe, 57, 58, 105, 106, 108
deflection angle, 96
delta function singularity, 95, 98
dimming of the signal, 97
Dirac delta function, 9, 105
directional singularities, 39
div-E equation, 62
div-H equation, 63
divergence of vorticity equation, 62
dragging of inertial frames, 84
dual (left, right) of Weyl tensor, 60

E
Eddington–Finkelstein coordinates, 28, 82, 105
electric part of Weyl tensor, 60
electromagnetic energy–momentum tensor, 36, 74
electromagnetic shock wave, 19, 23
Ellis–Bruni perturbation theory, 64, 78
energy conservation equation, 61, 62
energy–momentum–stress tensor, 60, 75, 78
entropy of the black hole, 113
equations of motion in first approximation, 38, 53
equations of motion of matter, 62, 65
event horizon, 81, 83, 119

expansion of null geodesics, 34, 43, 69, 87
external gravitational field, 28, 53
extra dimensions, 118
extrinsic curvature, 88, 122

F
Fermi–Walker transport, 133
fluctuating geometry, 110, 112
fluid 4-velocity, 75
formation of apparent horizon, 89
freedom of polarization, 12
future null-cones, 27, 29, 33
future null infinity, 81, 86, 113
future time-like infinity, 82

G
gauge-invariant cosmological perturbation, 64
gauge transformation, 11
Gaussian curvature, 54, 68
geodesic congruence, 98, 102
geometric optics approximation, 112
geometrical construction of a gravitational wave, 1
Gibbons–Penrose isoperimetric inequality, 91
gravitational collapse, 81, 87, 104
gravitational constant, 79, 118, 131
gravitational force, 118
gravitational potential barrier, 110, 115
gravitational shock wave, 23
gravitational wave, 24, 65, 76

H
half-null tetrad, 17, 22
Hawking radiation, 110, 113, 115–117
Hawking temperature, 113
head-on collision, 19, 21, 24, 103
head-on collision of photons, 91, 97, 99
Heaviside step function, 13
hierarchy problem, 118
high-energy, 119
high-frequency approximation, 26
high-frequency gravity wave, 25, 26, 97
high-speed Kerr black hole, 79
higher dimensional black holes, 126
Hodge dual, 19, 45
hoop conjecture, 87
hoop conjecture in D-dimensions, 123

I
impulsive gravitational wave, 9, 13, 20, 23, 95, 105
induced metric, 38, 59, 68
infinite blueshift, 104
infinitesimal Lorentz transformation, 6, 8, 45
information-loss, 109
ingoing null rays, 113
inhomogeneous cosmological models, 56
inner apparent horizon, 83, 104
inner horizon, 104
integral curve of a vector field, 11, 34, 56, 63
interior of a black hole, 105

intrinsic metric, 106
invariant null direction, 4
isotropic cosmological models, 56, 64, 67

K
Kerr black hole, 82, 86, 120, 125
Kerr particle, 39, 41, 44, 53
Kerr particle angular momentum, 41
Kerr–Schild form of line-element, 91, 93
Khan–Penrose solution, 10, 16
Killing horizon, 81, 85
Killing tensor, 85, 122
Killing vector, 81, 83, 85, 120
Kruskal null coordinates, 80

L
light-like boost, 91, 95, 125
light-like hypersurface, 106
limiting curvature principle, 105
line singularity, 95, 125
linear approximation, 10
Lorentz 3-force, 7
Lorentz boost, 2, 8, 91, 124, 127
Lorentz transformation, 2, 4–6, 8

M
magnetic part of Weyl tensor, 62
mass inflation, 104
mass parameter, 39
Maxwell 2-form, 19
Maxwell's equations, 19
mean energy flux of Hawking radiation, 115
metric fluctuations, 79, 110, 113
Minkowski inequality, 123
Minkowskian space–time, 68, 89
monochromatic wave, 24, 73, 103
multipole moments, 53, 94, 125
Myers–Perry metric, 120

N
naked singularity, 83, 87
Nariai–Bertotti space–time, 21
Newman–Penrose components, 9, 18, 20, 23, 24, 26, 132
no-hair theorem, 103
non-singular Lorentz transformation, 5
null-cone, 8, 89, 90
null generators, 87, 106
null hyperplane, 57
null hypersurface, 24, 65, 69, 83
null rotation, 4
null shell, 87, 105

O
optical coordinates, 102
orthogonal hyperplane projection, 123
orthonormal tetrad, 99, 133
outer apparent horizon, 87
Ozsváth–Robinson–Rózga space–time, 56

P

Papapetrou's equations of motion, 55
particle scattering, 96
past null infinity, 81, 113
past time-like infinity, 82
Penrose convex thin shell, 87
Penrose wave, 9
perturbed Einstein equations, 38, 48
perturbed event horizon, 112
perturbed half-null tetrad, 48
perturbed Maxwell equations, 37
perturbed Ricci tensor, 48
perturbed shear, 64
perturbed space–time, 29, 37
perturbed Weyl tensor, 67, 77
Petrov classification, 9, 23
photon scattering, 97
Planck length, 104
Planck mass, 118
Planck scale, 118
plane gravitational wave, 12
plane of rotation, 120
plane wave, 11, 24, 116
potential 1-form, 29, 34, 74
pp waves, 56
principal null direction, 11, 67
production of mini black holes, 118
profile of wave, 9, 13, 19, 25
proper, orthochronous Lorentz transformation, 2, 4

Q

quantum gravity, 104, 110

R

Raychaudhuri's equation, 62, 88
Reissner–Nordström black hole, 79
Reissner–Nordström particle, 33, 38
renormalized surface gravity, 113, 115
retarded time, 80, 127
Ricci identities, 61
Ricci tensor, 17, 19, 22, 26, 27, 31, 132
Riemann curvature tensor, 17, 20, 21, 58, 92, 93, 131
Riemann tensor centred approach, 96
Riemann–Lebesgue theorem, 25, 27
Riemannian connection 1-form, 131
Robertson–Walker space–time, 67
Robinson's limit of Schwarzschild, 5
Robinson–Trautman space–time, 27
rotating mass, 48
runaway motion, 29

S

scalar curvature, 18, 60
scalar massless field, 115
scattering properties of high-speed particles, 97, 124
Schwarzschild black hole, 79, 103
Schwarzschild coordinates, 79
semiclassical, 105, 110, 119

shear, 61, 99
shear propagation equation, 62
shear-free in optical sense, 69
singular hypersurface, 105
singular Lorentz transformation, 5
singular null hypersurface, 87, 105
singularity, 80, 83
spatial infinity, 81
spatial rotation, 2, 120, 133
specific angular momentum, 82
spherical harmonic, 35, 38, 51–55
spin tensor, 45, 52
spin vector, 39, 45
spin–spin terms, 41, 44
spindle singularity, 89
spinning test particle, 53, 55
static, 79, 81, 106
stationary, 79, 81, 83
stereographic coordinates, 68
surface energy density, 87, 122
surface gravity, 112
surface pressure, 106
surface stress–energy tensor, 87, 106

T

Tangherlini metric, 120
tetrad components of Maxwell field, 19, 34
tetrad components of Riemann tensor, 18
tetrad components of Weyl tensor, 35
thin shell, 87
time-like congruence, 61, 97
topology, 103, 119, 122
tortoise radial coordinate, 80
total mean curvature, 123
transplanckian, 110
transport law, 30, 133
trapped surface, 86, 123
twist, 43, 57, 87
twist-free, 33, 98

U

ultra-relativistic, 119
unimodular matrices corresponding to Lorentz transformation, 2, 6
uniqueness theorem, 119

V

vacuum fluctuation, 110
Vaidya metric, 110–112
violation of geodesic motion, 33, 41
vorticity propagation equation, 62
vorticity tensor, 61

W

wavefront, 11, 26, 38, 67, 74
weak field approximation, 125
Weyl conformal curvature tensor, 18, 36, 60, 72, 105, 132

Z

zero angular momentum observer, 83